9.50

GOVERNORS STATE UNIVERSITY LIBRARY
S0-ARM-092
3 1611 00342 6951

RADAR, SONAR, AND HOLOGRAPHY

AN INTRODUCTION

Radar, Sonar, and Holography

AN INTRODUCTION

Winston E. Kock
THE UNIVERSITY OF CINCINNATI
AND CONSULTANT, THE BENDIX CORPORATION

ACADEMIC PRESS New York and London 1973
A Subsidiary of Harcourt Brace Jovanovich, Publishers

ACADEMIC PRESS, INC.
111 Fifth Avenue, New York, New York 10003

United Kingdom Edition published by
ACADEMIC PRESS, INC. (LONDON) LTD.
24/28 Oval Road, London NW1

Library of Congress Cataloging in Publication Data

Kock, Winston E.
Radar, sonar, and holography.

1. Radar. 2. Sonar 3. Holography. I. Title.
TK6575.K62 621.3848 72-7695
ISBN 0-12-417450-7

PRINTED IN THE UNITED STATES OF AMERICA

To Gerrold R. Zacharias

CONTENTS

PREFACE

The purpose of this little book is to provide an introduction to the technology of radar and sonar. Because the new science of holography (for which its inventor, Dennis Gabor, received the 1971 Nobel Prize in Physics) is affecting both these fields quite strongly, the book includes an explanation of the fundamental principles underlying this new art (including the subjects of wave coherence, interference, and diffraction) and of the hologram process itself. Finally, numerous examples are discussed which show how holography is providing new horizons to radar and sonar systems. The book thus also provides a simple approach to the new technology of holography. It is therefore hoped that the discussions to follow will make clear the basic difference which exist between photography and holography, on the one hand, and, on the other, between standard sonar and radar, and the hologram versions of these two technologies.

The book should be useful as a supplementary reading assignment to early college students in science and engineering, and also to senior high students having a bent for science and engineering.

ACKNOWLEDGMENTS

The author dedicates this book to Professor Gerrold R. Zacharias of the Massachusetts Institute of Technology. In 1950, long after the end of World War II, Dr. Zacharias accepted an invitation from the U.S. Navy to be the director of a classified summer study called Project Hartwell. The esteem held for Professor Zacharias and the recognition of the value of this study was evident in the list of those who agreed to participate. It included, among others, an outstanding radio and radar authority, Dr. Harold T. Friis (the author's first supervisor at the Bell Telephone Laboratories), and the following persons with the positions they hold at the time of this writing: Dr. Jerome Wiesner (President, Massachusetts Institute of Technology), Dr. Edward E. David, Jr. (President Nixon's Science Advisor), Dr. James B. Frisk (President, Bell Telephone Laboratories), and Dr. Ivan Getting (President, Aerospace Corporation). Today many would find it surprising that a classified study, aimed at developing new avenues for further classified research, would have received such strong support as these names indicate.

The author is indebted to Allen H. Schooley of the Naval Research Laboratory for the ripple-tank photos (Chapters II and III), to F. K. Harvey of the Bell Telephone Laboratories for the photographs portraying microwaves and sound waves, to Lowel Rosen and John Rendiero of the NASA Electronics Research Center for various hologram photographs, and to The Bendix Corporation and the Bell Telephone Laboratories for various photos of radars and sonars.

The author also wishes to acknowledge that numerous figures herein and their related text appeared in various earlier publications of his, including articles in IEEE journals, the journals *Electronics, Applied Optics,* and the *Bendix Technical Journal,* and in his earlier books, "Sound Waves and Light Waves" (Doubleday, 1965), "Lasers and Holography" (Doubleday, 1969), "Applications of Holography" (Plenum, 1971), and "Seeing Sound" (Wiley, 1971).

This book is an expansion and updating of the author's presentation to the U.S.–Japan Seminar on Holography, held at Washington, D.C. on October 20–27, 1969, under the auspices of the National Academy of Sciences, the National Science Foundation, and the Japanese Society for the Promotion of Science. The author wishes to thank Professor George Stroke, State University of New York, and Head of the U.S. Delegation, for the invitation to participate as Co-chairman and as one of the ten U.S. delegates.

INTRODUCTION

The technology of radio has exhibited as rapid a growth, and found as wide a range of applications, as perhaps any single field of modern science. One of the more important branches of this technology, the radio image-forming process called radar, has, in recent years, benefited quite significantly from a new *optical* image-forming procedure called holography.

The transmission and reflection properties of radio waves were first conclusively demonstrated experimentally by the German scientist Heinrich Hertz in 1887. Nineteen years later, in 1906, the U.S. scientist Lee De Forest invented the three-electrode vacuum tube, a device destined to become a key factor in the mushrooming of technology to follow. The fields of radio communication and radio broadcasting, including television broadcasting, were the first to capitalize on the capabilities of radio transmission. That of radar was soon to follow. This field, utilizing the *reflection* of radio waves, and traceable even to Hertz's early experiments, is now equal in its importance and breadth of applications to radio communications and radio broadcasting.

Thus, radar, originally developed as a means for providing the military with information on the location and movements of enemy forces, has since found wide application in civilian sectors, sea transport, air traffic control (at airports), and aircraft navigation. This last use includes weather radars, with which aircraft pilots can detect and thereby avoid atmospherically turbulent thunderstorm regions in the air space. Even automobile travel is beginning to benefit from various applications of radar.

The importance of radar to the military in World War II led to a very rapid development of many of its capabilities, and also to very strict security measures. As early as February 1935, one of the more important documents on radar, submitted by the British scientist Sir Robert Watson-Watt to the British Air Ministry and entitled "Detection of Aircraft by Radio Methods," was classified "Secret."* Watson-Watt's proposal was successfully demonstrated during Royal Air Force maneuvers in 1937, and, following the very successful use of radar during the early years of World War II, Watson-Watt was knighted in 1942.

Sonar can be thought of as the acoustic form of radar. Its development actually preceeded radar, having been used by the U.S. Navy in the 1920s. However, the many important applications of radar, in both military and civil sectors, caused its development to proceed much more rapidly, so that today the sonar field often benefits from developments in the more recently developed technologies of radar. Accordingly, in recent years, sonar systems have also become highly sophisticated.

Both radar and sonar are echo-location systems, whereby the reflection of a small burst of energy (radio energy in the case of radar, and sound energy in the case of sonar) is detected, much as a handclap is heard as an echo from a distant vertical wall. The term radar was derived from the phrase "RAdio Detection And Ranging" and, following extensive use of the word radar, the word "sonar" was coined as a companion to it, being derived from the phrase "SOund NAvigation and Ranging."

Radar and sonar can be likened in many respects to photography, where light waves, rather than radio or sound waves, are employed to provide information about the position and properties of objects located in front of its viewing device, the camera. Accordingly, both radar and sonar have employed the lenses and reflectors (mirrors) of optical photography, and, for many years, were patterned much along the lines of classical photography.

* This author's first secrecy order was placed on one of his radar patent applications shortly after its submittal date of October 1941, more than a month before Pearl Harbor. At that early date the letter was simply labeled "Notice" although the body of the letter did state "you are hereby ordered . . . to keep the (subject matter) secret." Later letters, received by the author concerning other radar patent applications, were labeled not just "Notice," but "Secrecy Order."

In 1948 an event occurred which was soon to become quite significant in the history of modern radar and sonar. Again it was a British scientist who was responsible for the new development. In that year Dennis Gabor of the Imperial College in London described a technique called holography. Although the technique bears some striking resemblances to ordinary photography, it was quickly recognized to be a basically new process of wave recording. Gabor gave the photographic record involved in his procedure the name hologram, from the words "holo" meaning complete and "gram" meaning message. One of the spectacular features of a hologram is its ability to reproduce, or "reconstruct," its photographic record as a three-dimensional image, an image so realistic that it causes most viewers to reach out and try to touch the objects displayed.

Because this particular feature was so spectacular, the optical form of holography received the bulk of the attention during the initial period following its discovery. But soon radio and acoustic forms of holography were investigated, and exciting new forms of radar and sonar resulted. The U.S. scientist Emmett Leith was the first to apply, in 1963, the laser, a new light source first demonstrated in 1960, to optical holography, and because he and his colleagues at the University of Michigan, under the leadership of Dr. Louis Cutrona, were then working on a new form of radar, Gabor has pointed out that "Emmett Leith arrived at holography by a path just as adventurous as mine was," adding: "I came to it through the electron microscope, he through side-looking coherent radar."

In the latter portion of this book we shall see how holography has already helped radar and sonar find new performance horizons, such as in the synthetic-aperture radar, and how its use is being further extended to exotic new forms of radar and sonar signal processing, including an acoustical form of holography having certain advantages over standard sonar. A discussion of the basic principles involved in radar, sonar, and holography will be given, as well as a description of several rather recent forms of radar such as Doppler radars and phased array radars. Because a knowledge of certain properties of wave motion, including coherence and diffraction, are of importance in understanding the principles of many radar systems, a brief review of these properties will first be presented.

Chapter I

WAVE PROPERTIES

Because radar devices use radio waves and sonar devices use sound waves, a review of wave properties can help provide a better understanding of these systems.

Certain properties of wave motion are manifested by water waves such as those formed when a pebble is dropped on the surface of a still pond, as shown in Fig. I-1. Because all wave energy travels with a certain speed, such water waves move outward with a wave speed or *velocity of propagation* v. Waves also have a *wavelength*, the distance from crest to crest; it is usually designated, as shown, by the Greek letter λ. If, in Fig. I-1, we were to position one finger so that it just touched the crests of the waves, we could feel each crest as it passed by. If the successive crests are widely separated, they touch our finger less often, less frequently, than if the crests are close together. The expression *frequency* is therefore used to designate how frequently (how many times in one second) the crests pass a given point. The velocity, the wavelength, and the frequency (stated in *cycles per second*, or *hertz*, abbreviated Hz) are thus related by the expression:

frequency equals velocity divided by wavelength. (1)

Fig. I-1. Water waves in a pond, having a frequency f (the periodicity of their up and down motion), a wavelength λ (the distance from crest to crest), and a velocity of propagation v.

This says simply that the shorter the wavelength, the more frequently the wave crests pass a given point, and similarly, the higher the velocity, the more frequently the crests pass. No proportionality constant is needed in Eq. (1) if the same unit of length is used for both the wavelength and the velocity, and if the same unit of time (usually the second) is used for both the frequency and the velocity.

Coherence

The property referred to as the coherence of waves has assumed a great significance since the development of holography. This new technology demands "coherent waves," whether they are employed in purely optical applications (where the coherent light waves are generated by a laser), in radar applications (where highly coherent radio waves are involved), or in sonar or seismic applications (where coherent sound waves are used).

When waves are referred to as single-wavelength or, what is the

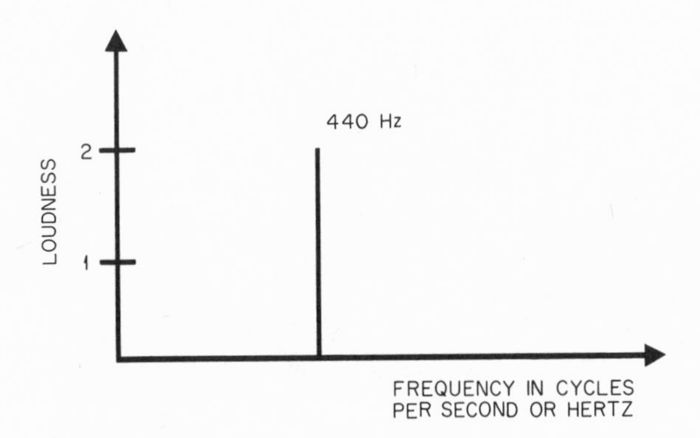

Fig. I-2. Representation of a single-frequency sound wave.

same thing, as single-frequency waves, they are said to have good *frequency coherence.* The sound from a tuning fork, the radio waves generated by a highly stable radio-wave source, and light from a laser all exhibit an extremely high degree of frequency coherence. One method of indicating the single-wavelength nature of a sound wave (the extent of its frequency coherence) is by the portrayal of the *frequency analysis* of the waves. In Fig. I-2, such an analysis is presented for a single-frequency sound wave. The frequency is plotted along the horizontal axis, and the amplitude or

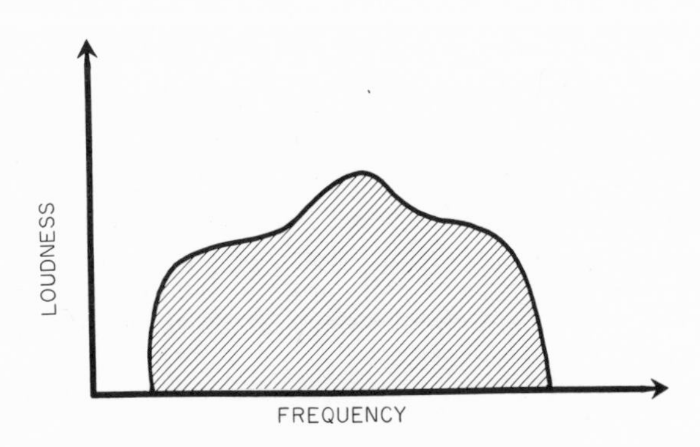

Fig. I-3. A noiselike or nonperiodic sound has a broad spectrum of frequencies.

loudness is specified by the height in the vertical direction. The frequency or pitch of the one-frequency component of this sound is indicated as 440 Hz (the note A in the musical scale), and its loudness as two (arbitrary) units. Such a tone would be classed as having an extremely high frequency coherence.

A sound having very little frequency coherence is the sound of noise. Noiselike sounds are very irregular and include, for example, the sound of a jet aircraft or the sound of howling wind in a storm; such sounds have a broad, continuous spectrum of frequencies. The analysis of a broad noiselike sound is sketched in Fig. I-3.

The sound portrayed in Fig. I-2 is that of a perfectly pure, single-frequency sound, in other words, one of infinitely high frequency coherence. Actually, such absolutely perfect waves do not exist in nature; however, the extent to which a wave approaches this perfection can be specified. Figure I-4 is a representation of a noiselike sound whose components extend over a fairly narrow frequency band. Its frequency coherence can be specified by stating the width of its frequency band as a percentage of the center fre-

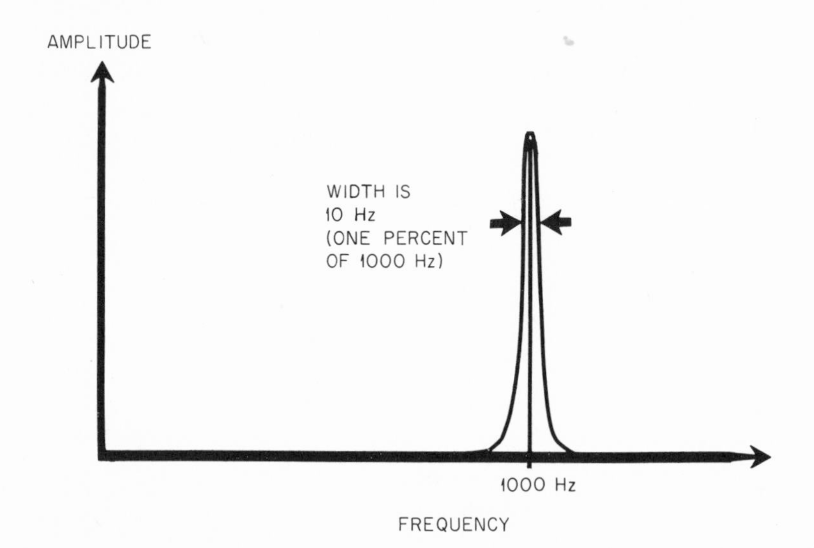

Fig. I-4. When a nonperiodic sound approaches in its nature a periodic one, its spectrum narrows.

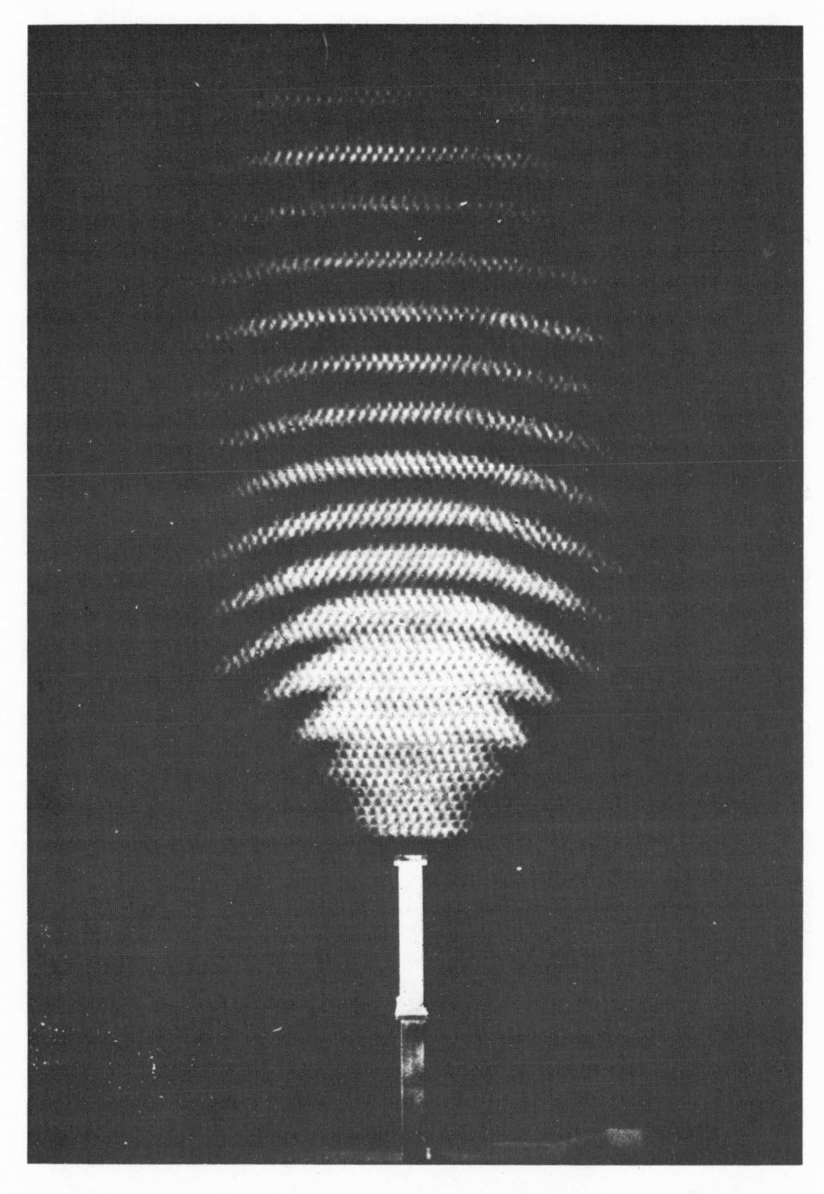

Fig. I-5. Single-wavelength microwaves issuing from a waveguide.

quency. In Fig. I-4, the bandwidth is 10 Hz, and the central frequency component is 1000 Hz. The sound thus has a 1% frequency spread, or a *bandwidth* of 1%.

Frequency Stability

Another way of specifying how closely a coherent wave approaches perfection is by stating its frequency constancy or frequency stability. The very fact that there is a frequency spread in the analysis of Fig. I-4 indicates that the frequency of the tone involved wavers back and forth in pitch over the frequency band shown. It is accordingly not absolutely constant in pitch, and the specification of a 1% bandwidth is equivalent to stating that the tone has a frequency instability (or stability) of one part in one hundred.

Electronic sound-wave generators (called *audio oscillators)* employ various procedures to achieve a high degree of pitch or frequency stability. Some audio oscillators achieve a frequency stability of one part in a million or better. For the application of holography in sonar or seismic exploration, such highly stable audio and subaudio sound sources are needed.

Radio waves are also usually generated electronically, and they too can be made very constant in frequency and, therefore, highly coherent. Figure I-5 portrays a set of highly coherent microwave radio waves issuing from the mouth of what is called a waveguide, a hollow rectangular tube connected to the microwave generator. * The need for extremely high frequency-stability oscillators in certain radio applications led to the development of an exceedingly narrowband radio device utilizing atomic processes to achieve its stability. It is called the *maser*, an acronym for *M*icrowave *A*mplification by *S*timulated *E*mission of *R*adiation. One variety, the hydrogen maser, can achieve a frequency constancy of one part in a million million. Following the development of the microwave maser, its principle was extended to the light-wave region, and light-wave "oscillators"

* The procedure for portraying wave patterns such as this will be discussed in Chapter VI.

having very high coherence then became available. These are called lasers (*L*ight *A*mplification by *S*timulated *E*mission of *R*adiation). One form of laser, the gas laser, exhibits a particularly high frequency constancy, and, as we shall see later, it is the one most often used for making holograms.

Spatial Coherence

A second form of coherence, that called *spatial coherence,* is shown in Fig. I-5. Here we see single-frequency waves having curved wave fronts. When such waves are passed through a lens such as the lens of Fig. I-6, they become plane waves. In this figure, sound waves having high frequency coherence are directed toward the lens at the right by the small acoustic horn shown at the left.

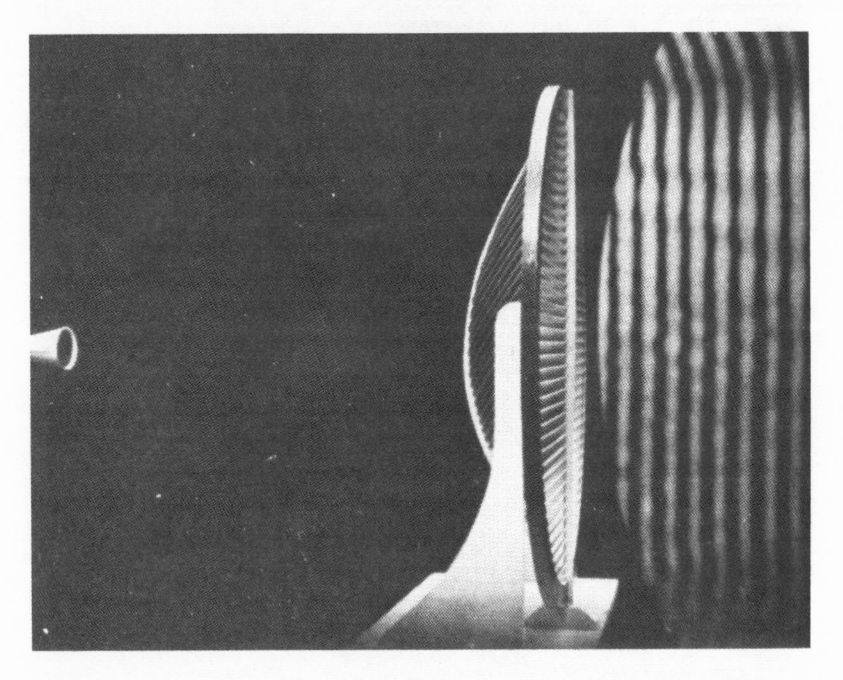

Fig. I-6. An acoustic lens converts into plane waves the circular wave fronts of sound waves issuing from a horn placed at the focal point of the lens.

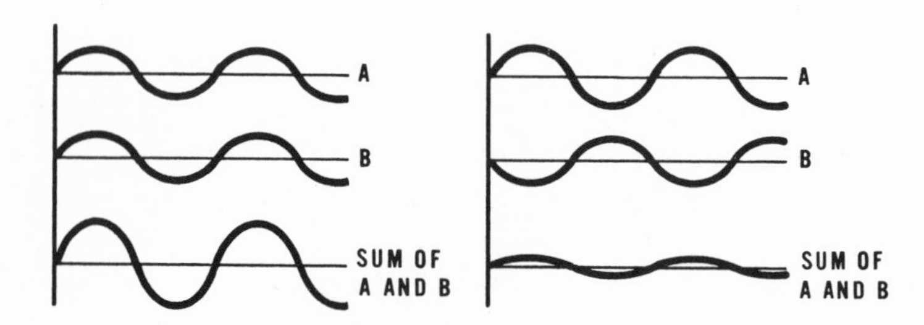

Fig. I-7. Two waves of the same wavelength add (at the left) if their crests and troughs coincide, and subtract (at the right) if the crest of one coincides with the trough of the other.

When they emerge at the far right, the planarity of the waves is evident. We shall see later that in holography this availability of uniform plane waves (that is, spatial coherence) is just as important as the availability of frequency coherence.

Wave Interference

Now, if a simple and uniform set of sound waves were to meet a second set of similarly uniform, single-wavelength waves, a phenomenon called *interference* would result. At certain points the two sets would add, a condition called *constructive* interference, and at others they would subtract, a condition called *destructive* interference. As sketched at the left of Fig. I-7, when the crests of one wave set A coincide with the crests of a second set B constructive interference occurs, and the height of the combined crests increases. When, on the other hand, the crests of one source coincide with the *troughs* of the second source, as shown on the right, destructive interference occurs, and the combined crest height is lowered. For sound waves, such additive and subtractive effects cause increases and decreases in loudness in the sound pattern; for light waves, they cause variations in brightness, or light intensity.

Phase

In discussing wave properties, the concept of *phase* is very useful. If we compare the position of the *crest* of one of the waves in

Fig. I-7 with the position of, say, one of the hands of a clock when that hand is pointing vertically upward (at twelve), and the bottommost part of the trough of that wave with a clock hand pointing vertically downward (at six), we see that we can relate the various portions of a wave to the 360° of a circle, as traversed by the clock hands. Thus, for the two top waves at the *right* of the figure, we can say that one is 180° displaced in *phase,* (or 180° *out-of-phase*) relative to the other. (The expression "of opposite phase" is also used.) The two top waves at the *left* of the figure, on the other hand, could be said to be "*in phase*" with each other. If, finally, the two waves were to have had a relative displacement halfway between the two conditions portrayed in the figure, one would express this situation by saying that the phase of one is displaced 90° from that of the other. Obviously much smaller differences in the relative phase of two waves can also exist, and we shall see later that in a "phased-array" radar the waves radiated from each of several individual or unit radiators, although identical in frequency, could be made to differ slightly in *phase,* thereby causing the "combination wave," as radiated by the totality of all the unit radiators, to be *aimed* in any one of many different directions, simply as a consequence of this phase difference between the unit radiator waves. Since the horizontal coordinate of Fig. I-7 represents the passing of time, a shift along that direction is a shift or delay in time. When the waves are not single frequency, being, say, like the noise signals discussed in connection with Fig. I-3, the concept of phase is not applicable, and then only the time-delay concept is useful.

Wave Diffraction

Webster's dictionary defines diffraction as "A modification which light undergoes, as in passing by the edges of opaque bodies or through narrow slits, in which the rays appear to be deflected, producing fringes of parallel light and dark or colored bands; also the analogous phenomenon in the case of sound, electricity, etc." From this definition we see that there are several aspects of diffraction. We shall discuss first the diffraction effects caused by one or more slits.

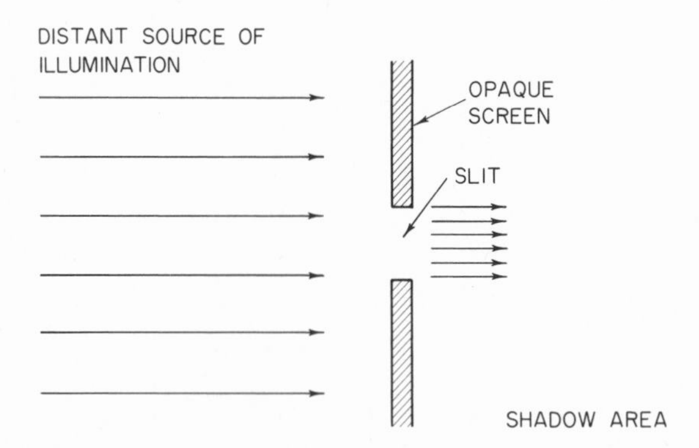

Fig. I-8. When a slit in an opaque structure is illuminated by a distant source of light, the energy distribution is uniform over the width of the slit.

Diffraction by a Slit

Let us consider the case of a distant light source or sound source illuminating a slit in an opaque screen, as shown in Fig. I-8. What is the radiation pattern that is formed by the slit in the dark area (the shadow area) behind the screen? In Fig. I-9 this illumination pattern, as cast on a second screen, is sketched. It has a bright central area flanked by a series of maxima and minima. The pattern is plotted in more detail in Fig. I-10. One property of this pattern, the width of the beam created at great distances by the waves passing through the slit, is rather interesting. This beamwidth is inversely

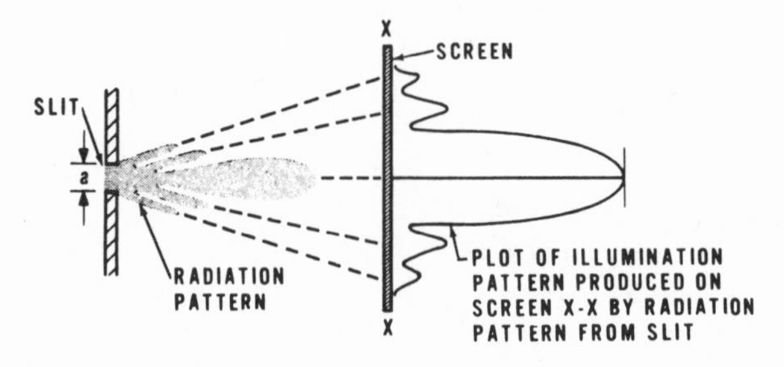

Fig. I-9. In the shadow region behind the opaque screen of Fig. I-8, light emerges from the slit in a beam-shaped pattern.

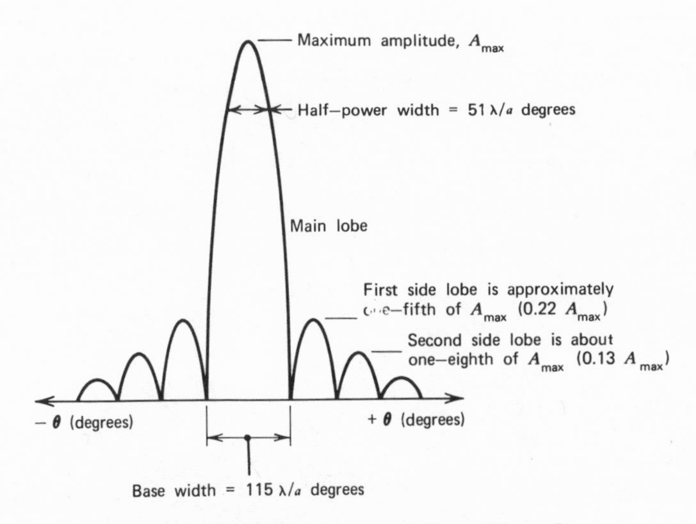

Fig. I-10. The diffraction pattern formed by a slit.

proportional to the aperture or slit width a and directly proportional to the wavelength λ; when a and λ are expressed in the same dimensional units, this width, in angular degrees, is $51\lambda/a$.

Let us see what sort of beamwidth this figure of $51\lambda/a$ corresponds to for the tubular paraboloidal reflector shown partially in Fig. I-11. This is a radar reflector for microwave radio waves of 3.4-cm wavelength employed in a World War II shipboard fire-control radar. The length of the horizontal aperture of this paraboloid of revolution is 8 ft, corresponding to approximately 244 cm. For fully uniform illumination occurring in a slit this wide when illuminated by waves of 3.4-cm wavelength, the beamwidth would be, in degrees, 51×3.4 divided by 244, or about 0.71°, which is the beamwidth one would observe at great distances from the slit.

The Near Field

We shall see later that in holography we are also interested in diffraction effects produced nearby rather than at great distances. Near the slit of Fig. I-8, the light pattern remains the same width as

Fig. I-11 An 8-ft parabolic reflector designed at the Bell Telephone Laboratories Radio Research Laboratory for a World War II shipboard radar. The person at the right peering into the lower cathode-ray tube is Harold Friis, to whom the author reported at the time.

the width of the slit; that is, the beam remains completely collimated in this region. The distance from the slit for which the beam dimension matches the aperture dimension is again a function of aperture size and wavelength. This distance is equal to the slit width or aperture dimension squared, divided by twice the wavelength. Thus the beam from an aperture 30 wavelengths across will itself remain 30 wavelengths across out to a distance 450 wavelengths from the aperture. Figure I-12 is a picture of sound waves being radiated from the lens of Fig. I-6 when placed in the mouth of a conical horn. Here the aperture is 30 wavelengths across. This photo portrays only the amplitude of the sound waves, and although the figure shows the beam only out to a distance of perhaps 30 or 40 wavelengths, it is seen that the edges of the pattern remain parallel, that is, the width of the beam remains the same as the aperture width for that distance.

Diffraction by Two Slits

We now consider the diffraction effects that occur when light passes through two slits in an opaque screen. Numerous very bright and very dark areas result, and these are called, in optical terminology, *fringes*. The two-slit effect of optics can be simulated with two nondirectional coherent sound sources as shown in Fig. I-13. This figure portrays the fringe pattern formed by two sources separated by three wavelengths; in this photo the wavelength of the sound waves is approximately 1.5 in. The combination of two identical wave fields results in wave addition and wave cancellation. As would be expected, wave addition occurs at points equidistant from the two sources, that is, along the center line of the two radiating points; this is the central, bright, horizontal area. Wave cancellation (destructive interference) occurs at those points where the distance from one source differs from that to the other source by one-half

Fig. I-12. The sound-amplitude pattern directly in front of a thirty-wavelength aperture radiator.

Fig. I-13. Two separated sound sources act like two optical slits in creating a diffraction pattern by constructive and destructive interference.

wavelength. At such points, one of the two wave sets has crests (positive pressures) where the other has troughs (negative pressures). The two areas where this half-wavelength destructive interference effect is evident in Fig. I-13 are the black areas immediately above and below the central bright area.

Bright areas are again seen above and below these two black areas. These are areas where the distances from the two sources differ by one *full* wavelength. One of the wave sets is a full wavelength ahead of the other and wave crests and wave troughs again coincide, so that positive pressures add to form higher crests and negative pressures add to form deeper troughs. Similarly, in those areas where one of the two wave sets is two wavelengths ahead of the other, wave addition again results; these are the shorter bright areas at the very top and very bottom of the figure.

Dependence of Diffraction on Wavelength

The diffraction pattern of Fig. I-13 was formed with single-frequency sound waves of a particular frequency (wavelength). Different wavelength waves would have produced a different pattern. This effect is sketched in Fig. I-14. Waves having the longer wavelength λ_1 are one wavelength behind in the direction *A*, whereas those of wavelength λ_2 are one wavelength behind in a different direction.

Diffraction by a Knife-Edge

We turn now to the second kind of diffraction included in Webster's definition, diffraction caused by the edges of opaque bodies. When waves pass by such edges, some energy is deflected (diffracted) into their shadow region. This phenomenon is illustrated in Fig. I-15. In this photograph, sound waves are arriving from the left,

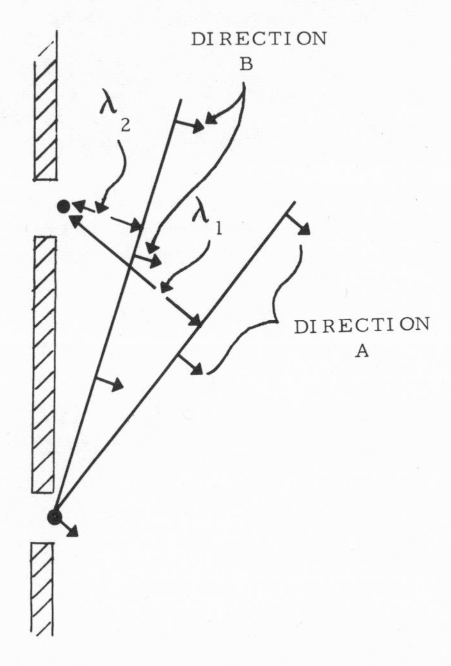

Fig. I-14. The diffraction angle for waves emerging from two slits or from a grating depends on the wavelength. Because holograms are a form of grating, their performance also depends upon wavelength.

Fig. I-15. Plane sound waves arriving from the left proceed unhindered at the top of the photo. In the shadow region, fainter circular wave fronts are evident, caused by diffraction at the edge of the shadowing object.

and a long wooden rectangular board acts as a shadowing object. In the region above the board, the waves are seen to continue unimpeded toward the right as plane waves. Because the waves cannot penetrate the board (it is opaque to the sound waves), the lower right of the photo is a shadow region. The edge of the board, usually referred to as a knife-edge, is seen to act as a new source of sound energy. The faint waves in the shadow region have wave fronts that are circular (cylindrical), and the common axis of these cylinders is located at the knife-edge.

Diffraction by a Disk

If the opaque object is, instead of a knife-edge, a disk, energy is diffracted from all points on the perimeter of the disk, and a much more complicated pattern exists in its shadow. This wave pattern is shown in Fig. I-16. We see from the knife-edge diffraction of Fig. I-15 how sound diffracting around the disk edges can form this pattern. In Fig. I-16 sound waves are also arriving from the left, and in the upper and lower parts of the photo, in the nonshadow areas, the waves proceed unhindered as in Fig. I-15, as parallel wave fronts. In the shadow region itself, two sets of circular wave fronts are evident, one having the top edge of the disk as its center of curvature, and the other having the lower edge as its center. In the figure, circular wave fronts are arriving from all points on the perimeter of the disk; these points all act as new sources of wave energy, and these many wave sets all interfere with one another. This combination establishes a narrow wave pattern along the axis, a pattern that looks much like the pattern of the unimpeded waves at the top and

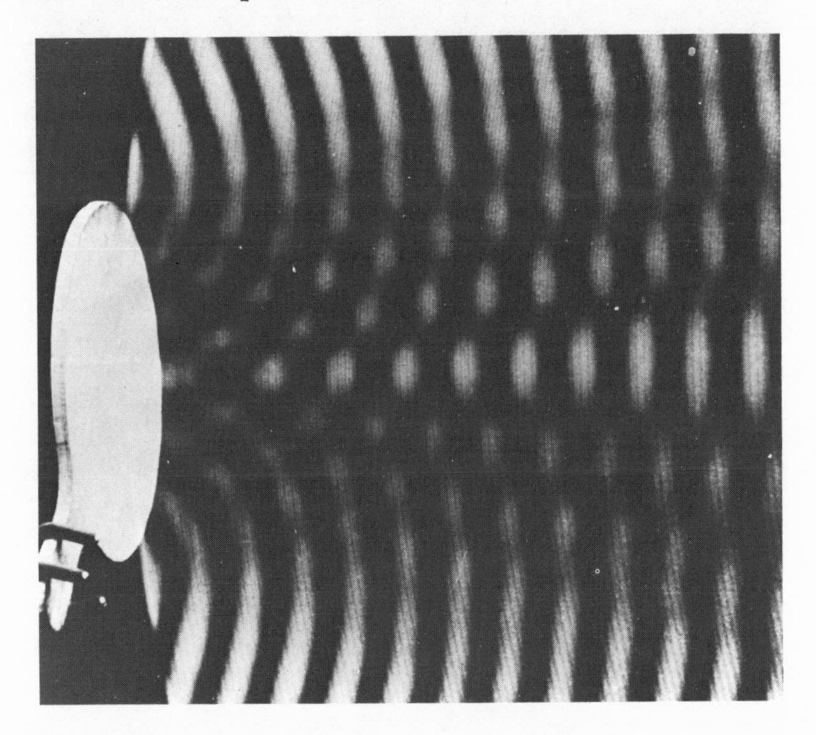

Fig. I-16. Sound waves diffracted around a circular disk combine in the shadow region and produce a central beam of parallel wave fronts.

Fig. I-17. When the amplitude pattern in the shadow of a disk is examined, the bright central lobe is clearly evident.

bottom of the photo. Thus, the combination of the many new wave sources positioned along the circular perimeter of the disk produces a concentration of wave energy along the axis.

Lord Rayleigh observed this effect for both light waves and sound waves. Figure I-17 omits the wave fronts of Fig. I-16 and portrays only the amplitude pattern of sound intensity behind the disk. For these two figures, the same disk and same wavelength sound waves were used. In this figure we see clearly what Lord Rayleigh described as a bright central spot or cone, and also the "revival of effect" above and below the "ring of silence."

Disks and Zone Plates

Rayleigh noted that the concentration of sound waves in the shadow of a disk can be further enhanced by positioning additional annular rings around the disk, thereby blocking out other areas or zones of wave energy. This process of adding additional blocking rings leads to a zone plate, a device used for focusing several forms of wave energy. We shall see later that a zone plate is closely related to a hologram.

Diffraction by a Zone Plate

The zone plate can be described as a set of flat, concentric, annular rings that diffract wave energy. The open spaces permit passage of waves which add constructively at a desired focal point, and the

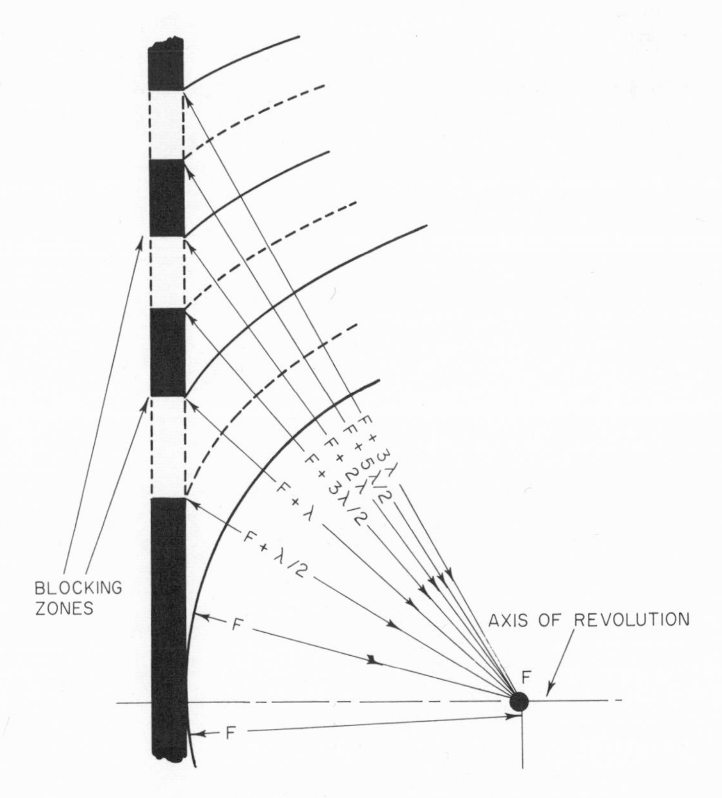

Fig. I-18. When the diffraction of a disk is supplemented by the added diffraction of annular rings, a zone plate results.

opaque rings prevent passage of waves which would interfere destructively with these waves at that point.

Figure I-18 indicates the procedure for determining the positions of the rings in a zone plate have a circular opaque disk for its central portion. The positions of the blocking rings are determined by drawing circles whose centers coincide with the desired focal point and whose successive radii from this focal point differ from one another by one-half of the design wavelength. Thus, at that point where the first one-half wavelength circle intersects the plane of the zone plate, the central blocking zone, the opaque disk, is terminated. Farther out, at the one wavelength point, the first annular blocking ring is started. It is terminated at the one and one-half wavelength point, and the process continues. Inasmuch as the design of a zone plate is based on one particular wavelength, waves having wavelengths that differ from this design wavelength will not be affected by it in the desired way.

Chapter II

WAVE RADIATORS

In their operation, both radars and sonars radiate and receive short pulses of wave energy. A key element, therefore, in their performance is the radiator employed. In the case of radar, these radiators are designed to radiate microwaves, whereas in the case of sonar, sound waves are radiated. The radiating device is usually made directional, that is, it is made to aim its energy in a given direction so that targets located at different angles can be separately detected. In this chapter, we shall review ways in which directionality can be imparted to sound-wave and microwave radiators.

Dipoles

A familiar sight in the early days of television was two rigid horizontal wires supported at their center point, which constituted the simplest television receiving antenna. (In Great Britain, vertically oriented radio waves were used for television and the British television antennas, therefore, had their wires positioned vertically.) This simplest form of radio antenna is called a dipole. It is pictured in Fig. II-1. The two input leads are connected to the center of the dipole as

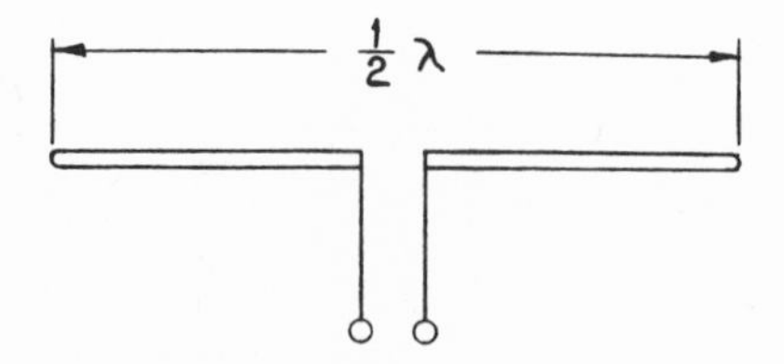

Fig. II-1. An electromagnetic dipole.

shown, and the dipole radiates most effectively when each of its halves is made to be a quarter wavelength long.

Unfortunately, this simple dipole is not very directional. (Its energy is radiated in many directions when used as a transmitting antenna, and it similarly receives energy from many directions when used as a receiving antenna.) Accordingly, present television anten-

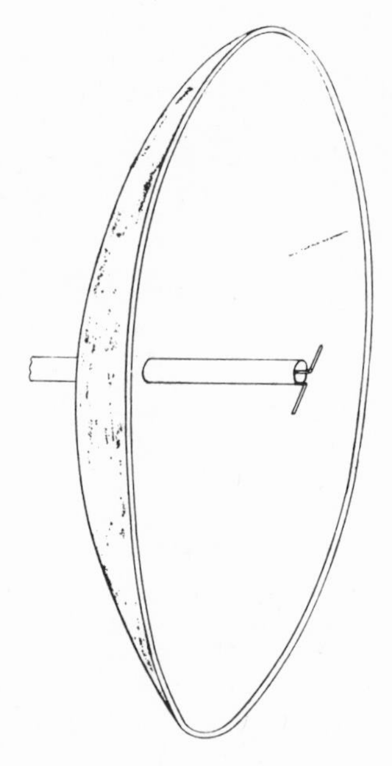

Fig. II-2. An electromagnetic dipole radiator acting as the feed for a parabolic-reflector radar antenna.

nas are generally more complicated than earlier ones; they consist of many groups of wires that permit the received signal to be made more intense at the receiving point. In radar, the dipole, if used at all, is generally combined with a larger structure such as a parabolic "dish" as shown in Fig. II-2. The dish reflects and beams energy that otherwise would have continued on to the left of the figure.

Dipoles are peculiar to radar-wave technology; sound waves, being a pressure disturbance, are not benefited by radiating structures that resemble the electrical dipole. However, in the field of microwaves, the use of a new transmission device called a waveguide led to the development of an improved form of radiator very closely resembling a device known throughout the ages to be able to

Fig. II-3. The nondirectional pattern of sound waves issuing from a telephone receiver.

enhance the directivity of sound waves. The device in question is the horn, and both electromagnetic waves emanating from a waveguide and sound waves emanating from a tube can be directed more effectively by the addition of a horn.

Horns

We are all familiar with the way a megaphone can cause sound radiation to be enhanced in a given direction. The radiation pattern of short-wavelength radio waves (called microwaves) issuing from a *small-aperture* radiator (a waveguide) shown in Fig. I-5, and the radiation pattern of sound waves issuing from a telephone receiver (Fig. II-3) both show no concentration or directional effect. The sound pattern of an acoustic horn (Fig. II-4), on the other hand, shows that the energy is concentrated in the direction in which the horn is aimed. As in the case of the dipole and dish reflector of Fig. II-2, it is the larger *aperture* of the horn that is responsible for its directionality. Because all the energy is confined within the horn, the intensity at its mouth remains quite high. In addition, the wave

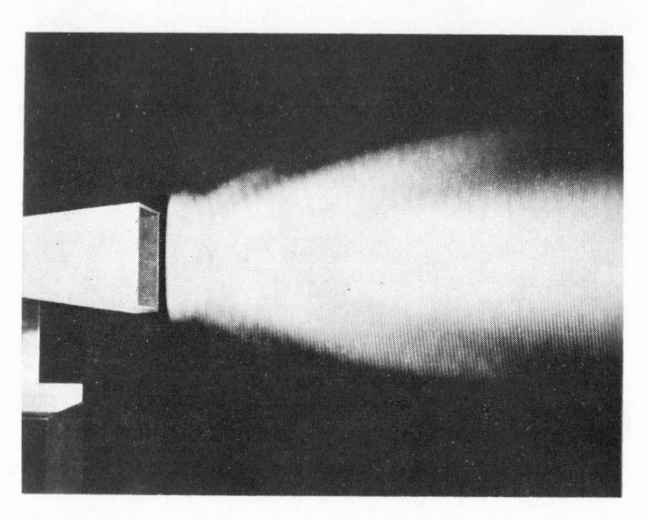

Fig. II-4. The directional pattern of sound issuing from a pyramidal horn.

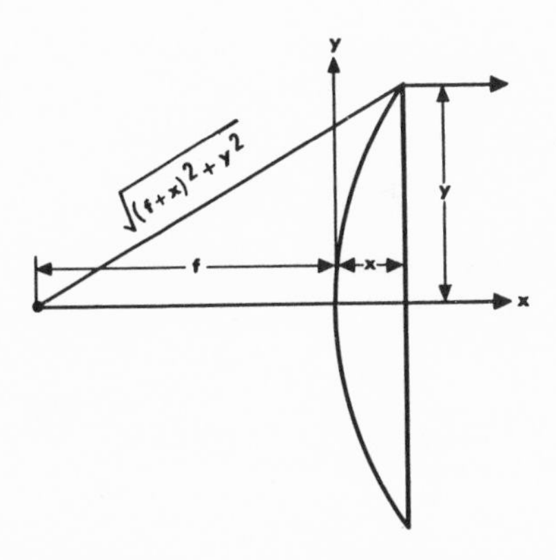

Fig. II-5. For the radiated wave fronts from a lens to be flat, the times of travel of all rays from the focal point to the flat front surface of the lens must be equal. The lower velocity of propagation within the lens accomplishes this if the lens is made thick at its center. For the phase length of the two rays to be equal $f/V_0 + x/V = [(f + x)^2 + y^2]^{1/2}/V_0$ or, if $V_0/V = n$, $(n^2 - 1)\ x^2 + 2fx\ (n - 1) - y^2 = 0$. The equation of the lens contour is a hyperbola.

fronts at the mouth of the horn are rather flat, and since energy propagates along a line perpendicular to the wave fronts, a concentration of sound energy results in the direction in which the horn is pointed.

Lenses

In Fig. I-6, we saw how circular wave fronts of sound waves can be made plane by the use of an acoustic lens. Dielectric lenses used for radio waves and light waves are also thick in the center and thin at the edge (that is, they are similar in shape to the acoustic lens of Fig. I-6). The lens profile is defined more explicitly in Fig. II-5. For energy originating at the focal point, a ray with a part of its path lying within the refracting material of the lens should be equal in "phase length" or "delay time" to a ray not passing through the lens. That is to say, the time taken for wave energy to arrive at the front face of the lens, that is, the amount of time delay, should be the same for all parts of the lens. To meet these conditions, the con-

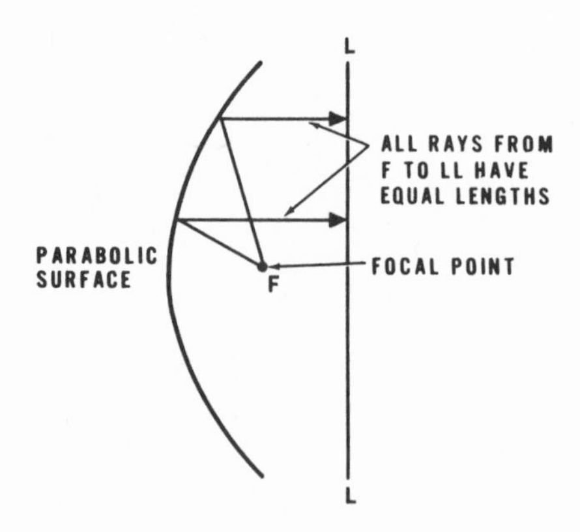

Fig. II-6. A curved reflector also can create the situation where the times of travel of all rays from a focal point to a plane are alike.

tour of the lens must be that of a hyperbola. For the profile to be hyperbolic in all planes, the surface will be a hyperboloid of revolution. In practice, most optical lenses have spherical surfaces because such surfaces are easier to fabricate than hyperboloidal ones to the accuracy required at the extremely short wavelengths of light. For the much longer wavelengths of radio and sound waves used in radar and sonar, this problem does not exist, and hyperboloidal surfaces are usually employed.

Parabolic Reflectors

Another form of wave radiator that converts circular wave fronts into flat wave fronts in a small space is the focusing reflector (Fig. II-6), equivalent to the dish of Fig. II-2. When the curve of that figure is a parabola, energy originating at the focal point *F* will, after reflection from the curved surface, arrive at the line *LL* with flat wave fronts. Figure II-7 shows a ripple-tank photo in which circular water waves are converted to plane waves by a parabolic reflector. A reflector that is a paraboloid of revolution will effect focusing in all planes. Parabolic reflectors are used in searchlight mirrors, radar antennas, and in some acoustical microphones designed for listening to distant sounds.

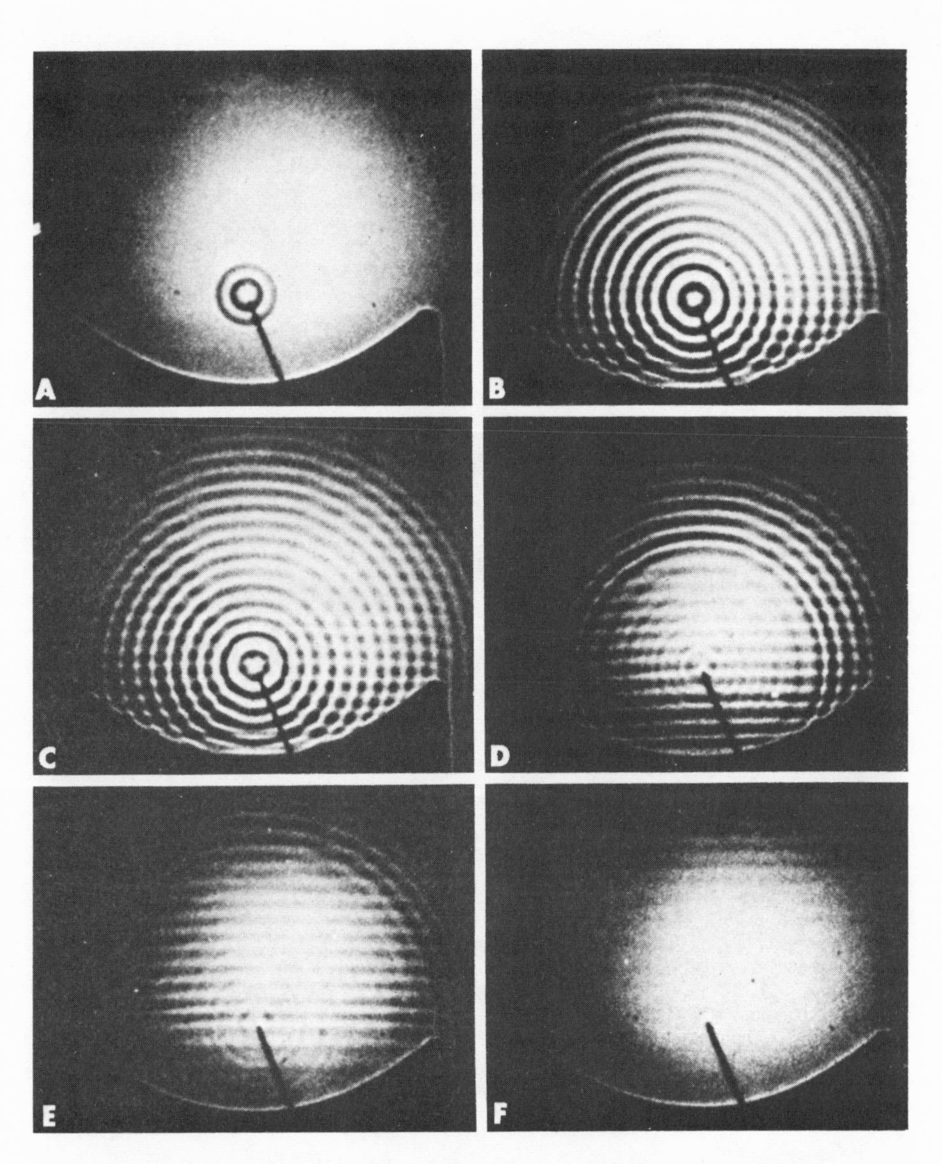

Fig. II-7. Photos of a ripple tank showing a parabolic reflector converting circular waves into plane waves (courtesy of U.S. Naval Research Laboratory).

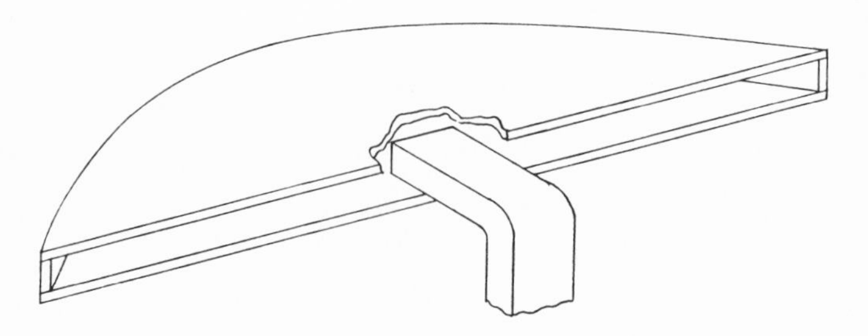

Fig. II-8. A pill-box or "cheese" parabolic antenna, having an open-ended waveguide as its feed.

We noted earlier that the radiator feeding the parabolic dish of Fig. II-2 is a dipole. It terminates a tubular conductor called a coaxial cable, in which the two leads are coaxial, one being the central conductor and the other the tubular outer conductor.

Another type of parabolic antenna extensively used in radar is the parallel-plate or "pill-box" antenna shown in Fig. II-8. (A similar form was called a "cheese" antenna by the British.) Here the mouth of a waveguide is placed between two parallel, conducting plates, and the circular wave fronts, propagating between the plates, are directed toward the *cylindrical* parabolic reflector. There con-

Fig. II-9. A pill-box antenna having horn flaps to provide it with greater directivity in the vertical plane.

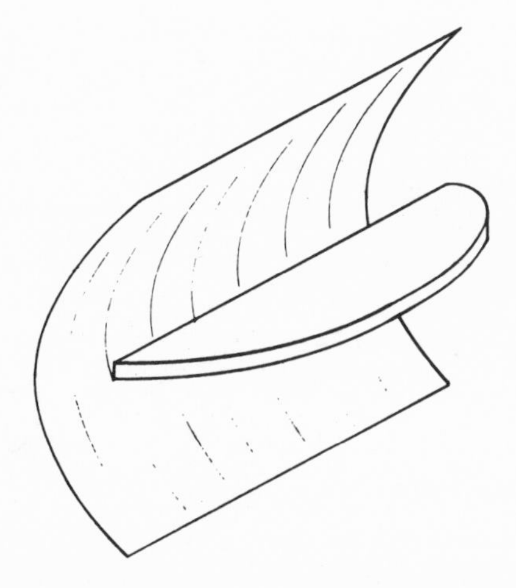

Fig. II-10. A pill-box antenna feeding a cylindrical parabolic reflector antenna.

verted to plane waves, the energy is reflected and emerges from the line aperture of the parallel plate structure. The long length of the line aperture provides a narrow horizontal beam, and its small height causes the vertical beam to be quite broad.

One method for narrowing the vertical beam of a pill-box antenna is shown in Fig. II-9. Here the plates are extended to form a hornlike structure, and the larger vertical aperture thereby generated provides the sharper vertical beamwidth. This arrangement is one commonly used today in marine radars; we shall discuss these in more detail in Chapter III. Another procedure is shown in Fig. II-10. Here a second, quite large cylindrical parabola is illuminated by the line aperture of the pill-box antenna. The sharp horizontal beamwidth is unaffected, whereas the circular (cylindrical) wave fronts generated in the vertical plane by the line aperture are, upon reflection from the parabolic cylinder, converted into plane waves. The reflected waves are thus plane in both vertical and horizontal directions. To avoid having the pill box in the central path of the final reflected waves, a variation of this procedure has occasionally been employed, as shown in Fig. II-11. Here only half of the reflec-

Fig. II-11. A pill-box antenna feeding only the upper half of the cylindrical reflector of Fig. II-10.

tor of Fig. II-10 is used, so that the reflected waves pass *over* the pill-box antenna feed.

Waveguides

In the years just prior to and during World War II, a new method of transmitting electromagnetic energy was developed. This method, which evolved mainly from the work of research groups at the Massachusetts Institute of Technology and the Bell Telephone Laboratories, used hollow tubes called waveguides to conduct the very short-wavelength radio waves (microwaves). The radiator of the microwaves shown in Fig. I-6 is a rectangular waveguide.

Figure II-12 is a ripple-tank photo of water waves, simulating the case of microwaves propagating in a waveguide. The channel causes the water waves to have a phase velocity that is dependent on wavelength, just as the phase velocity of microwaves in a waveguide is dependent on wavelength. In the left photo the circular pattern shows fairly long-wavelength waves radiating from a point source and then entering the waveguide structure. For this long-wavelength case, the "guide wavelength" at the right is seen to be

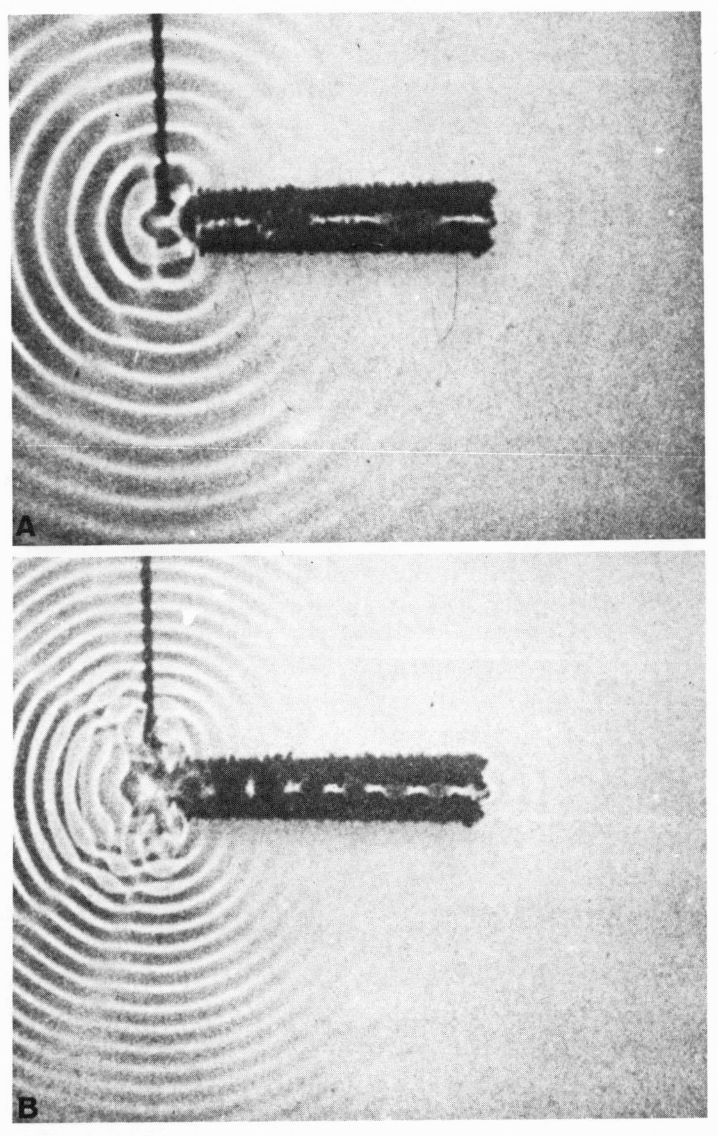

Fig. II-12. Photos of a ripple tank showing the increased phase velocity (longer wavelength) of waves confined in a waveguide (courtesy of U. S. Naval Research Laboratory).

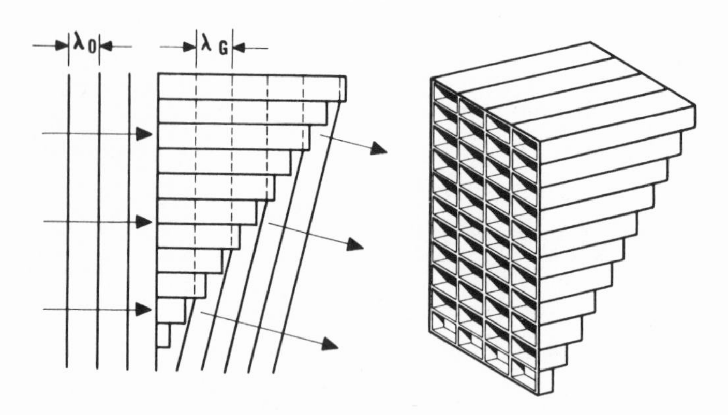

Fig. II-13. A wedge-shaped stack of waveguides acts as a prism because of the higher wave velocity within the waveguides.

much longer than the wavelengths of the circular pattern. In the photo at the right, the wavelength has been made shorter (as is evident from the circular pattern), and the *change* in the wavelength within the guide relative to the circular pattern wavelength is seen to be much smaller than the change shown in the photo on the left.

Waveguide Lenses

In a few instances lenses have found application in radar. Solid dielectric lenses, such as the one sketched in Fig. II-5, generally proved too heavy for the large antenna sizes desired. Thus, a 10-ft diameter lens made of polystyrene would weigh several tons, far more than the equivalent aluminum reflector. Accordingly, lighter-weight refracting structures were developed for use with microwaves, and microwave designers then could take advantage of certain properties of lenses.

The most widely used lens "material" utilized the higher phase-velocity property of waveguides just discussed. We saw in Fig. II-12 that within a waveguide the wave crests travel faster than in free space (actually, the *energy* travels more slowly). Since the process of focusing modifies the wave fronts, it is the higher wave-crest velocity that is instrumental when waveguide effects are used for lenses.

To see how waveguides can be used to refract microwaves, let us imagine a stack of rectangular waveguides cut to proper length and placed side by side, as shown in Fig. II-13. Let us assume that waves with flat wave fronts arrive from the left and emerge at the right. We see in the figure that the higher wave velocity within the guide (indicated in the figure by the longer guide wavelength λ_g) will impart a tilt to the wave front as the waves leave the assemblage. The structure thus behaves, for microwaves, like a prism, and this "refracting" property of waveguides has been used widely in the construction of lenses of various types and sizes.

In a rectangular waveguide the existence of the higher phase velocity is not dependent upon the presence of the top and bottom walls of the guide. Waveguide lenses, therefore, are usually made of sheets of metal, as shown in Fig. II-14. This lens is 18 in. in diameter, and it was constructed during World War II. It was, to the author's best knowledge, the first two-dimensional waveguide lens. A

Fig. II-14. A two-dimensional metal-plate (waveguide) microwave lens (built during World War II).

Fig. II-15. Another view of the lens of Fig. II-14, showing a horn-lens feed.

second view of it, with a smaller lens acting as the directional feed for the lens, is shown in Fig. II-15. Successful tests of this structure encouraged further exploration of numerous variations. The illustration brings out the fact that lenses, making use of the increased wave-velocity property of waveguides, must be thick at the edge and thin at the center, the exact converse of a glass lens. Figure II-16 shows a water wave ripple pattern that simulates the wave focusing effect of a waveguide lens. As was shown in the ripple-tank pattern of Fig. II-12, water waves propagating in a simulated radio waveguide situation have their wavelengths increased in length. This is equivalent to an increase in their velocity (phase velocity). The figure shows that the circular waves (originating at the center of the circles) become flat after passing through the waveguide lens.

The top and bottom portions of the flat waves at the right traveled along the longest paths within the high-velocity lens section, and this was responsible for the conversion to plane waves.

Stepped Lenses

To reduce further the weight of waveguide metal-plate lenses, the process called "stepping" was developed. In an unstepped circular lens, starting at the center of the lens and moving outward toward the rim, the thickness increases continuously. In a stepped lens, "setbacks," where the lens thickness is reduced abruptly, are incorporated. Because the step design is based on one given wavelength, such lenses are feasible only in those systems that employ frequencies lying close to the design frequency. Proceeding outward from the center of the circular lens of Fig. II-17, when the thickness in-

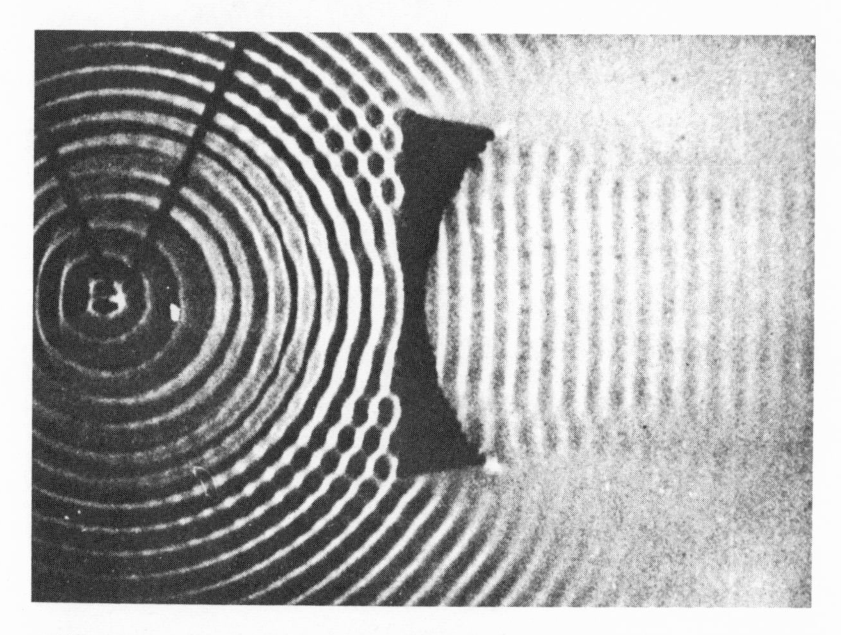

Fig. II-16. A ripple-tank photo showing the focusing action of a waveguide lens (courtesy of the Naval Research Laboratory).

Fig. II-17. A stepped circular waveguide lens antenna designed for shipboard radar use.

crease equals one design wavelength as measured within the lens—that is, one guide wavelength, the lens thickness can be reduced to its original center thickness. At succeeding points, wherever the thickness increase again equals a guide wavelength, succeeding steps are introduced. Figure II-18 shows the (unstepped) *front* surface of this lens; this lens was designed for use with a shipborne radar. Figure II-19 portrays a lens that has been stepped in the vertical direction only. It, like the cylindrical reflectors of Figs. II-10 and II-11, is called a cylindrical lens, inasmuch as it focuses only in one plane; that is, it focuses cylindrical waves emanating from a line source, as in Fig. II-10, rather than spherical waves emanating from a point source, as in the case of the lens of Fig. II-17. This lens was incorporated in a World War II mortar-locating radar. Two line sources such as the pill-box parabola of Fig. II-10 were placed one above the other, causing the combination of these with the lens to generate *two* pencil beams. The construction of the line sources was such that *their* contribution to the beam position (its horizontal, or azimuth location) could be varied (scanned). This resulted in two almost horizontal planes, one pointed in a higher direction than the other, being continually under observance (by virtue of the horizontal or azimuth scanning process. Whenever a mortar shell, on the upward portion of its trajectory, passed through these two planes, its azimuth position in each was ascertained, and a similar determination was made on the downward portion of its trajectory. Since a shell trajectory is completely defined when a

Fig. II-18. The front face of the lens of Fig. II-17 showing the absence of steps on this face.

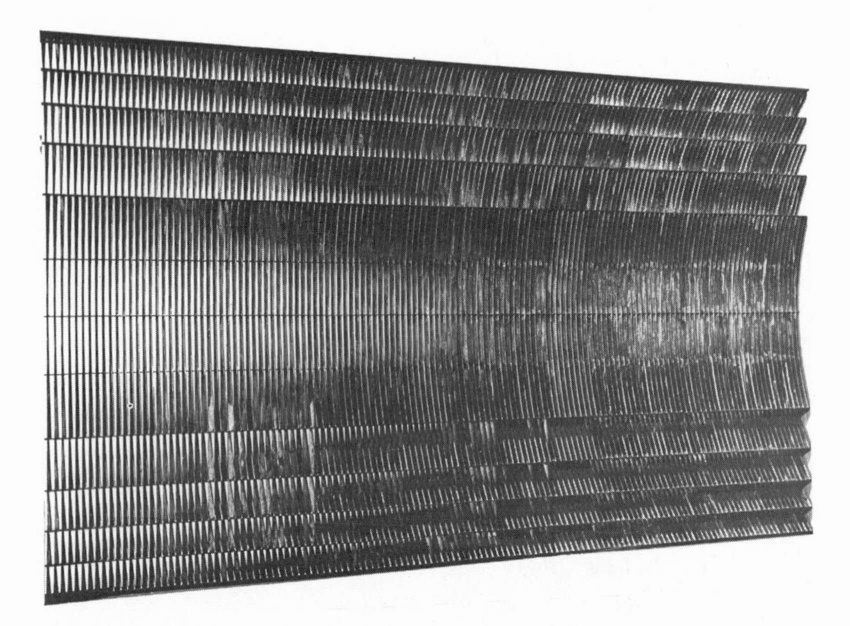

Fig. II-19. A stepped cylindrical lens designed for a mortar-locating radar.

number of points on it are specified, the location of the origin of the mortar shell (relative to location of the radar) could quickly be calculated from this radar information, and action to silence the mortar could then quickly be taken. A field model of this radar, designated the TPQ-2 (it was also fondly called "TPQ-2 and Tyler, too") was in production in December, 1944. The same waveguide principles underlying the design of the microwave lenses just discussed can be applied to the design of lenses for underwater sound sonar applications. Here the metal plates are replaced by parallel, completely resilient plates. H. Nodtvedt of Norway has described experiments with a 32-cm lens of 30 kHz, noting that this construction "makes it possible to construct acoustic lenses in liquids."

Arrays

We have seen how the horn, the lens, and the parabolic reflector obtain radiated wave fronts having flat phase. Still another method of achieving this goal is to use a large number of individual radiators, all energized in the proper phase. The simplest method for doing this is illustrated in Fig. II-20, where many small sound sources are connected together, and with the phases of all radiated signals made exactly alike. The individual radiators themselves are nondirectional, but when they are placed in a plane their outputs combine in such a way as to create two plane waves moving in opposite directions. If, as shown in Fig. II-21, a reflecting plate is placed the proper distance behind the plane of the radiators, the wave moving in that direction will be reflected and will add to the wave moving in the opposite direction. An array can thus provide very high directivity and yet have a very small thickness dimension.

The spacing of the individual radiators is important. If we place the radiators far apart (as the two sources are in Fig. I-13), objectionable side lobes are produced. Theory shows that if the radiators are spaced a half wavelength apart, phase addition can occur only in the direction perpendicular to the line joining the two radiators, and objectionable side lobes are eliminated.

Radiation perpendicular to the plane of the radiators is called "broadside radiation." In most broadside arrays the individual radiators are spaced a half wavelength apart.

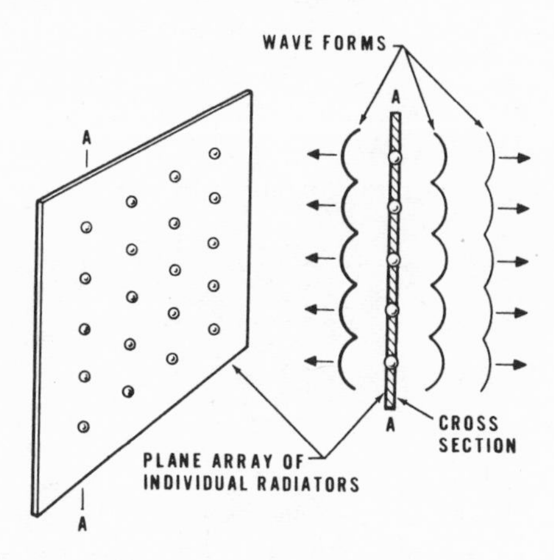

Fig. II-20. An array of individual radiators, themselves nondirectional, becomes a directional radiator of plane wave fronts if the elements radiate in phase.

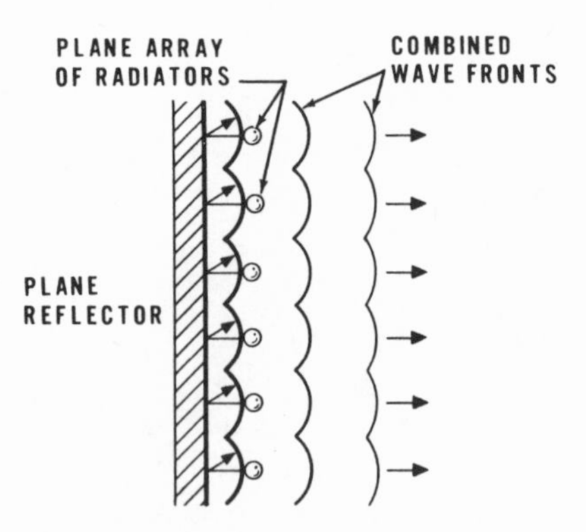

Fig. II-21. A reflector plate placed behind the array of Fig. II-20 causes it to radiate in one direction only.

End-Fire Arrays

Figure II-22 shows a series of radiators energized by a single source but with a delay incorporated between the individual radiating elements. If the individual delays are made equal to the time of radiated wave travel from one element to the next, the radiated waves will add in phase along the line joining the radiating elements. Such arrays are called "end-fire arrays." They radiate off the end of the linear structure making up the array. The array directs wave energy as a hose directs water.

Individual radiators can be energized by a transmission line (Fig. II-22), or the transmission line itself can be used as the radiator.

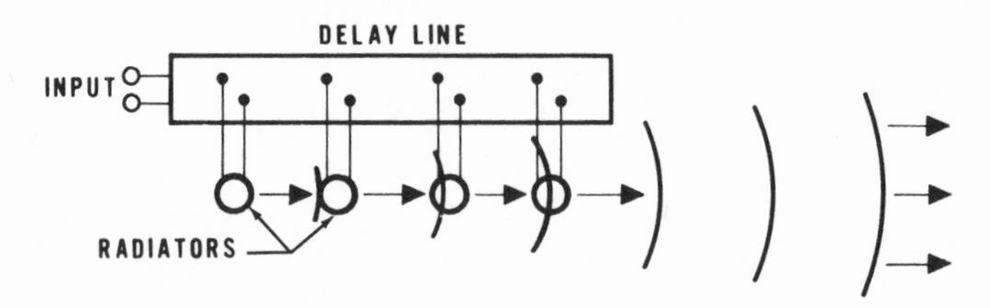

Fig. II-22. An end-fire array results when radiators are placed in a row and energized, not in phase, but in such a way that proper phase addition occurs in a given direction. In the illustration, this direction is to the right.

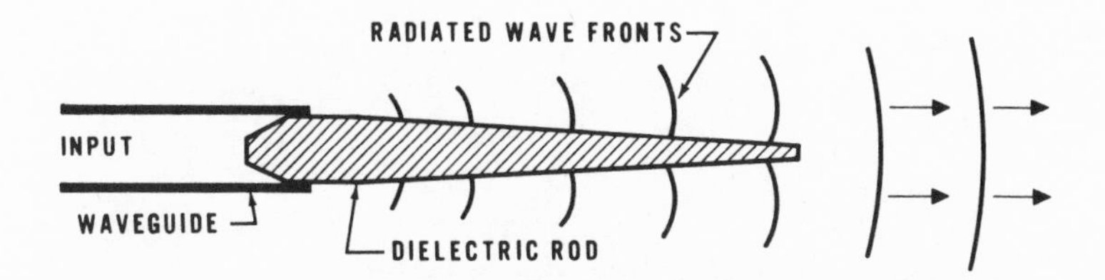

Fig. II-23. A tapered dielectric rod, which not only guides electromagnetic energy but also radiates it gradually, constitutes a useful end-fire radiator.

Figure II-23 shows a microwave end-fire radiator used in a shipboard fire-control radar. Microwave energy arriving from the left is confined within a hollow waveguide, and into the end of this tubular guide there is inserted an insulating (dielectric) rod. Because the dielectric rod employed was made of the translucent plastic called polystyrene, it was given the name "polyrod." The waves continue to advance into the rod, but all along the rod wave energy "leaks" out. The rod is tapered in its cross-sectional area, and, as the waves progress, more and more energy continues to be radiated from it over its entire length. Because the energy radiating from successive parts of the dielectric rod is traveling at a velocity very close to the velocity of light in free space, the dielectric rod acts like the end-fire radiator we have described. Good directional patterns are achieved in the direction in which the dielectric rod is pointed.

An acoustic end-fire radiator, patterned after Fig. II-23 and designed for underwater sound use, was described by P. Wille of Goettingen University.

Steerable Broadside Arrays

If varying amounts of delay are introduced between the energy source and the radiating elements of the arrays of Fig. II-20 or II-21, the beam, instead of being perpendicular to the plane of the array, can be "steered," that is, pointed in other directions. Thus, if the top element of Fig. II-20 is left unchanged, and a small section of the delay line of Fig. II-22 is inserted between the source and the next lowest element, and then still more delay line introduced for each of the succeeding elements proceeding downward, the emerging beam will be aimed downward. This is because energy will emerge first from the top radiator and successively later from the others, causing a tilt in the beam pattern.

Figure II-24 shows an acoustic prism causing this delaying effect. The velocity of propagation for sound waves in the prism material is less than that in free space, so the lower, thicker portion acts like the delay lines discussed above, and the otherwise horizontal beam is tilted downward.

In radars and sonars, this ability to steer the beam electrically, that is, without mechanically moving the radiator, permits much larger arrays to be used. These arrays are called phased-arrays and will be discussed in more detail in Chapter V.

The very large, several story-high radar of Fig. II-25 certainly does not lend itself to beam steering by mechanically rotating the structure. However, because it is equipped so that all of its thousands of radiating elements can be properly "phased," that is, energized with proper amounts of delay, its beam can be steered quite

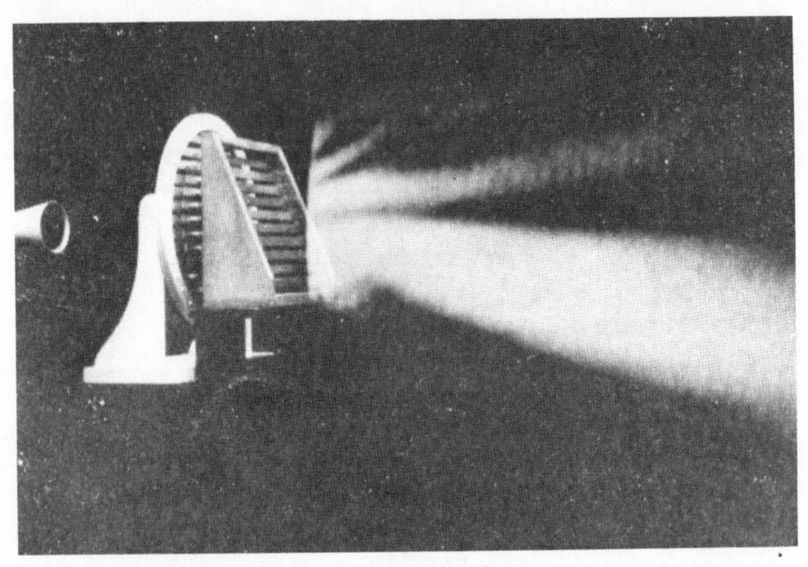

Fig. II-24. The focused beam of sound waves from an acoustic lens is deflected downward by an acoustic prism.

Fig. II-25. A multi-story radar shown under construction for the U.S. Air Force by the Bendix Corporation.

effectively by electronic means. This radar, called the AN/FPS-85 Spacetrack Radar, or the "Spadat radar" for short, was built by the Bendix Corporation. It is a high-power phased-array radar capable of automatically detecting, tracking, and cataloging hundreds of space objects simultaneously.

Gain

It is useful to be able to designate how effectively radiators concentrate their energy in a beam pattern. The term *gain* is used to indicate the improvement over a radiator that has no beam effect. A radiator that radiates equally in all directions is called an *isotropic* radiator. The gain of a broadside radiator is proportional to its area and inversely proportional to the square of the wavelength. Analytically, the ideal gain is expressed as $G = 4\pi A/\lambda^2$. When, as is usually the case, the antenna is not perfectly efficient, the A in this formula is called the "effective" area (smaller than the actual area). A high-efficiency square radiator 100 wavelengths on a side can beam energy to a distant point about 120,000 times more effectively than an isotropic radiator. This is the same as saying that if the power in the square radiator were to be reduced by a factor of 120,000, the radiated signal still would be received, at a distant point, as well as from an isotropic radiator transmitting the original high, unreduced power.

Since "gain" is just as effective in receiving as in radiating or transmitting, a radar or sonar, using the same radiator for transmitting and receiving, benefits from the *product* of the gains of the radiating antenna and the receiving antenna.

Tapered Illumination

As we saw in Fig. I-10, strong side lobes exist in the pattern of a uniformly energized or illuminated slit. For radar and sonar, a

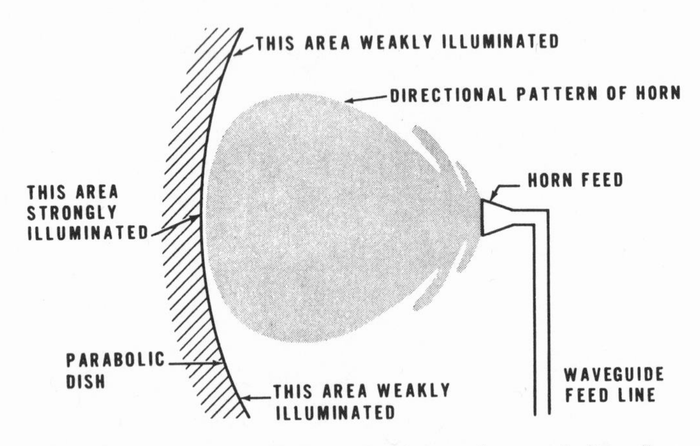

Fig. II-26. A small horn placed at the focus of a parabolic reflector produces a tapered illumination because of the directive effect present even in the small horn.

Fig. II-27. A metal lens designed to provide a special beam pattern.

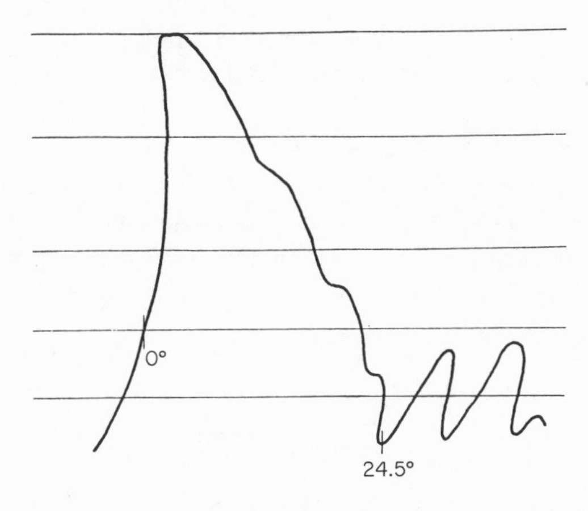

Fig. II-28. The pattern of the lens of Fig. II-27 as recorded on an automatic plotter.

process known as illumination tapering is employed to suppress the side lobes. For example, in radar applications when a parabolic *dish* is used as a radar antenna, the waveguide at its focal point is often equipped with a horn whose directivity brings about tapered illumination. The horn causes the raidated energy to be concentrated more at the center of the dish than at the edges (Fig. II-26), and the side-lobe level is thereby reduced.

Tapers have one disadvantage; they cause a lowering of the gain of the aperture. Maximum gain requires uniform illumination. However, because side-lobe level is a rather important consideration, gains of about half of this maximum value are usually acceptable in radar applications. Tapers also cause a beam broadening, resulting in the beamwidth formula of Fig. I-10 having the factor $51\lambda/d$ changed to approximately $65\lambda/d$ for the usual tapered illumination.

Cosecant-Squared Patterns

Often it is desired that the directional pattern of a radar be made to conform to a shape that differs from the usual "single-main-lobe" pattern. Thus, airport radars occasionally require that aircraft be detected not only at long ranges (so that the single beam would then point at the horizon) but also when they are very near the airport (then the single beam would have to be pointed upward at the plane). One very desirable form of an all-purpose airport radar is called a "cosecant-squared" beam, since it provides gain not only at the great horizon distances, but also at the higher-angle close-in points. A small metal (waveguide) lens antenna that forms such a beam is shown in Fig. II-27. The shape of the plates was deter-

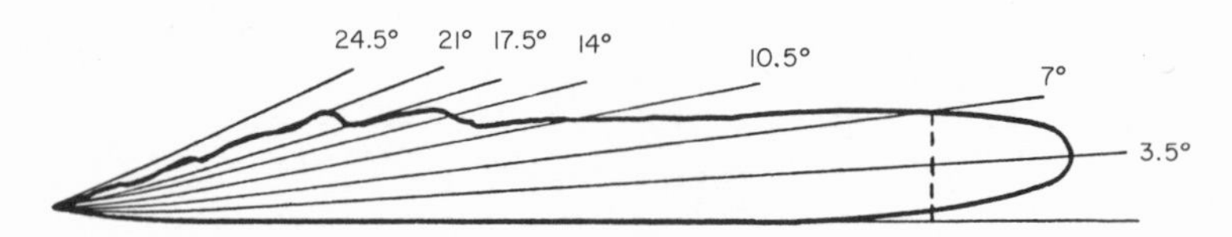

Fig. II-29. The plot of Fig. II-28 converted to the actual vertical beam pattern of the antenna.

mined from calculations based on a theory presented during World War II by U.S. scientists L. J. Chu and R. C. Spencer at the Massachusetts Institute of Technology. Figure II-28 shows a test pattern of this antenna as taken on an automatic pattern plotter, and Fig. II-29 shows the same pattern plotted in the horizontal direction. It is seen in these two last figures that at angles lower than 3.5°, where the gain is not needed, the pattern drops rapidly. At angles higher than 3.5° the pattern falls off gradually, retaining some "gain" even at angles as great as 21°. Because the planes are at shorter ranges at these angles, the gain, as seen, is made smaller (than it is, for example at 3.5°).

Chapter III

FUNDAMENTAL CONCEPTS IN ECHO-LOCATION SYSTEMS

Active Systems

In the most common form of radars and sonars, short pulses of energy are radiated periodically, usually from a highly directional radiator, and immediately after each pulse is transmitted the radiator is made to act as a receiver. If a reflecting object is located in the beam of the radiator, the energy in the pulse will strike it and will be scattered in many directions. Some of the energy, however, will be reflected back toward the radiator (now acting as a receiver) as a greatly weakened but still short pulse of energy. This received pulse is amplified and displayed to the radar or sonar operator.

Figure III-1 portrays four different instants occurring in a water-wave (ripple-tank) arrangement that simulates the action of a radar. At the upper left, a pulse of energy is generated, corresponding to the radar pulse and producing the waves shown. At the upper right the waves are starting to move toward a model of an aircraft. In the next photo the waves have reached the target and some reflection has started to take place. Finally, in the fourth photo, the reflected energy has returned to the vicinity of the transmitter, which now could be acting as a receiver, thus notifying an operator of the presence of a target. It is obvious that this photo is not to scale.

Because the velocity of propagation of radio waves and sound waves is known, the distance or range of the reflecting object can be determined from the length of time taken by the echoed pulse in returning to the receiver. Thus, for radio waves, which travel at the speed of light (186,000 miles/sec), a reflecting object at a 9.3-mile range (corresponding to an 18.6-mile round-trip path) would generate a returned pulse at the receiver one ten thousandth of a second after the transmitted pulse was radiated. Objects at other ranges would generate received pulses at different times, so that a record of all objects at all ranges which lie within the transmit–receive beam can be obtained.

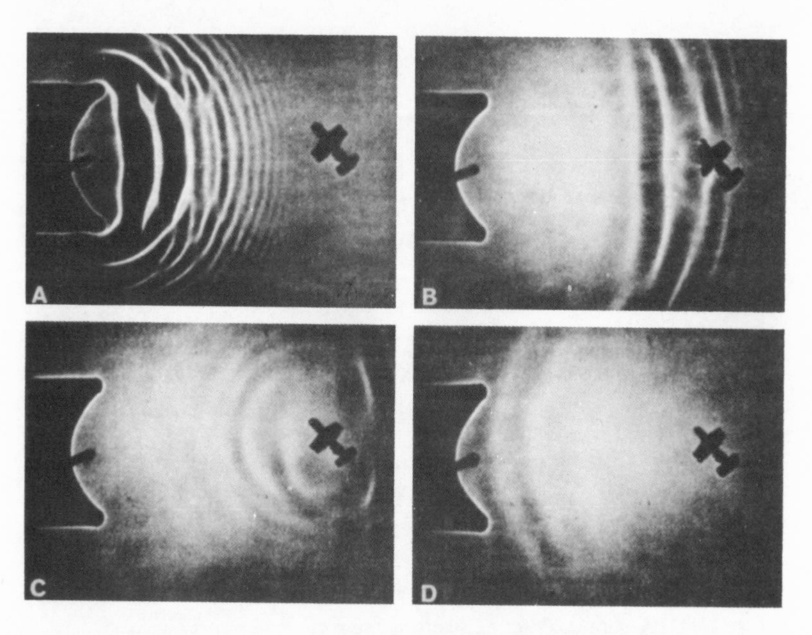

Fig. III-1. Ripple-tank photos show a pulse of wave energy being sent out and reflected back by a model (courtesy of U.S. Naval Research Laboratory).

If now the direction in which the beam is pointed is altered slightly, other objects, lying in the new beam direction, will generate "echoes," and the range of these can also be made available to the operator.

Figure III-2 is a sketch of the display available in a typical marine radar. It represents the face of an electronic device called a cathode-ray tube. In this device, which is very similar to the well-known "picture tube" of a television set, a very narrow, movable beam of electrons is directed toward the rear face of the circular

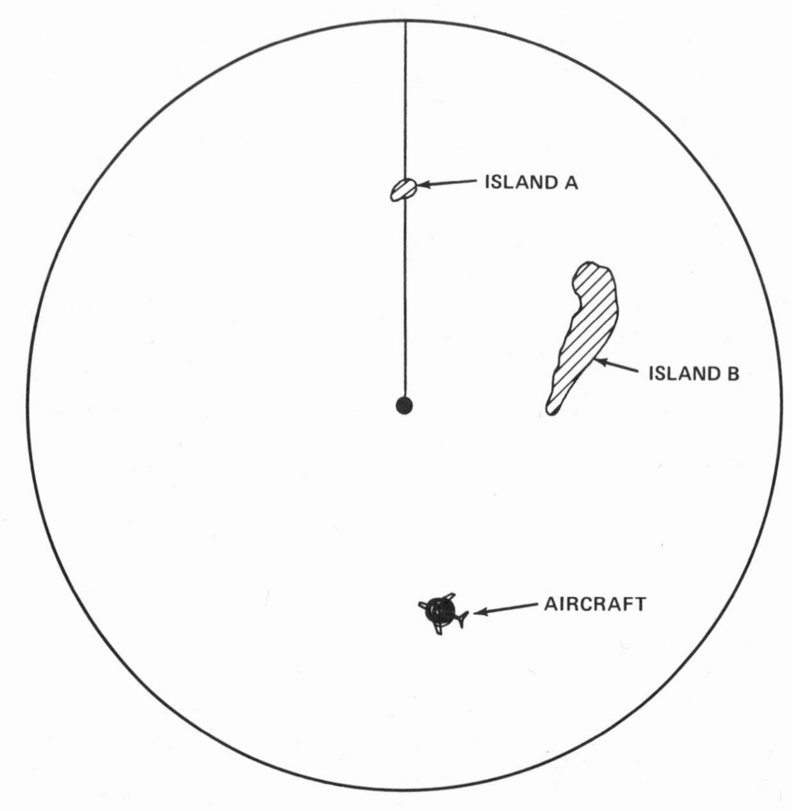

Fig. III-2. A plan-position indicator display as employed on a marine (shipboard) radar. The ship carrying the radar is located at the central dot.

glass plate of the figure and this face is coated with an electrosensitive phosphor that glows when struck by the electron beam. The vertical line in the figure corresponds to one sweep of the electron beam, as it proceeds from the center of the circle upward toward the edge. The strength of the electron beam, and therefore the brightness of the visual signal it generates on the phosphor, is controlled by the strength of the echoes as received and amplified. Thus, along the vertical line shown, only one echoing area exists, the island labeled *A*. The speed of motion of the electron beam is made to match the range coverage desired. Thus, if the maximum range desired is say 20 miles (so that island *A* might be at 12-mile range from the ship carrying this marine radar) the speed of the electron beam is made to match the echo travel time corresponding to an echo at 20-mile range.

The beam is now pointed in a new direction, say slightly to the right of the straight line, and a new pulse is transmitted. Another set of echoes results; again, in this case, it is only from island *A*. As the beam continues to move clockwise, it eventually covers the entire circle and starts over again. This form of display is called a "plan-position indicator." The phosphor in the cathode-ray tube of such a radar display is made to have some "persistence" so that the entire picture dies away only gradually. In the sketch a second island *B* is indicated, and a smaller object (really a dot) corresponds to an aircraft. For this latter echo, a sizeable movement of this *dot* will be noticeable in each new complete radar picture, whereas the island will have a motion (relative motion) corresponding to the slower speed of the ship.

It is evident that the detail which can be incorporated in the displayed "map" will depend upon: (1) the sharpness of the transmit–receive beam (this determines the angular or azimuth resolution), and (2) the pulse length (this determines the range resolution).

Active *sonar* comprises equipment in which bursts of *sound* are generated and transmitted in the hope of receiving an answering echo. Thus, a destroyer uses active sonar to echo range on a possible submarine contact. Active sonar is easily applied to give both range and bearing of targets within range.

Echo-ranging sonar of the "searchlight" type was developed a decade or more before World War II. This type of sonar projects intense pulses of sound along a relatively narrow beam (for example, 15–20°). If the beam encounters the target, an echo returns to the projector and gives an audible "beep" or a visual indication of its presence. The distance of the target from the searching vessel can be determined from the time required for the pulse to go out and the echo to return. The limitation of this arrangement is the considerable time required to cover the bearings of interest by successively training the projector, sending out a "ping," and listening for an echo. This is a consequence of the low speed of sound in water (approximately 5000 ft/sec).

Because electromagnetic waves do not propagate in sea water (it is a conductor of electricity), sonar has proved to be the only reliable means for searching for submerged objects such as submarines.

Pulse Considerations

Figure III-3 portrays two pulses of different length as used in sonar systems. The one at the left is rather short, lasting for only 10 msec. The one on the right has a length of 100 msec. (The term ping is often given to sonar pulses since they often sound like the word "ping" when they are received by the sonar after reflection.)

Figure III-4 shows the change in character of the echoes ob-

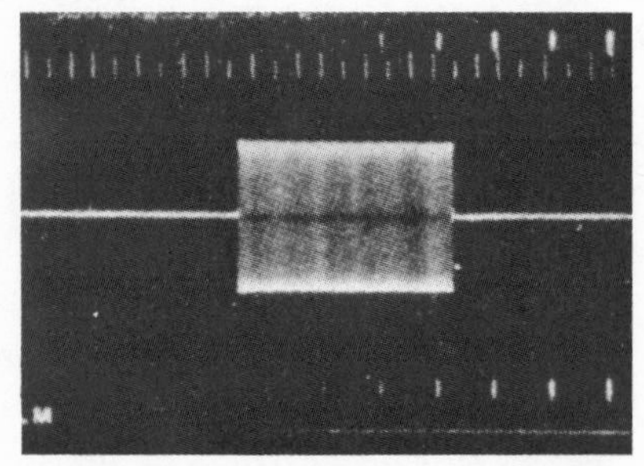

Fig. III-3. A short (10-msec) sonar pulse (left) and a longer one (100 msec) at the right.

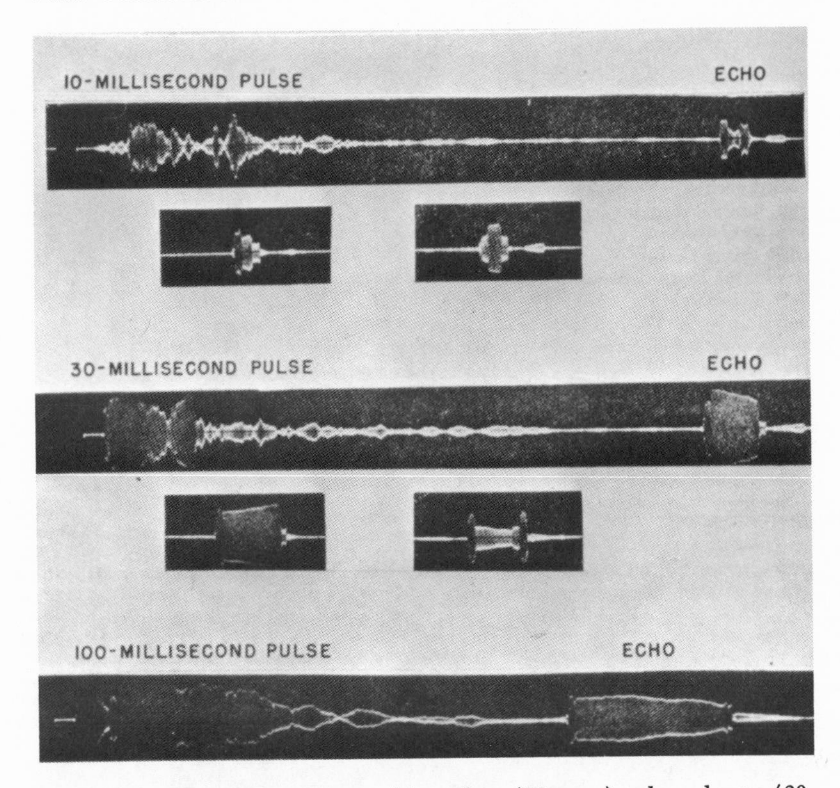

Fig. III-4. Sonar echoes generated by a short (10-msec) pulse, a longer (30-msec) one, and a still longer (100-msec) one.

tained when sonar pulses of different lengths are employed. In the top line, a 10-msec pulse and its echo are portrayed. In the middle line, the pulse length has been increased to 30 msec. It is seen that the echo is now larger and, of course, longer. The lower line portrays the situation for a 100-msec pulse. Here the echo is quite long and well defined. Had the usual noise that always exists been superimposed on these records, one can understand that the top echo would have been difficult to detect. Accordingly, the choice of pulse length involves a compromise. The shorter the pulse, the more accurate the range information on the target. Conversely, the longer the pulse, the greater the detection capability.

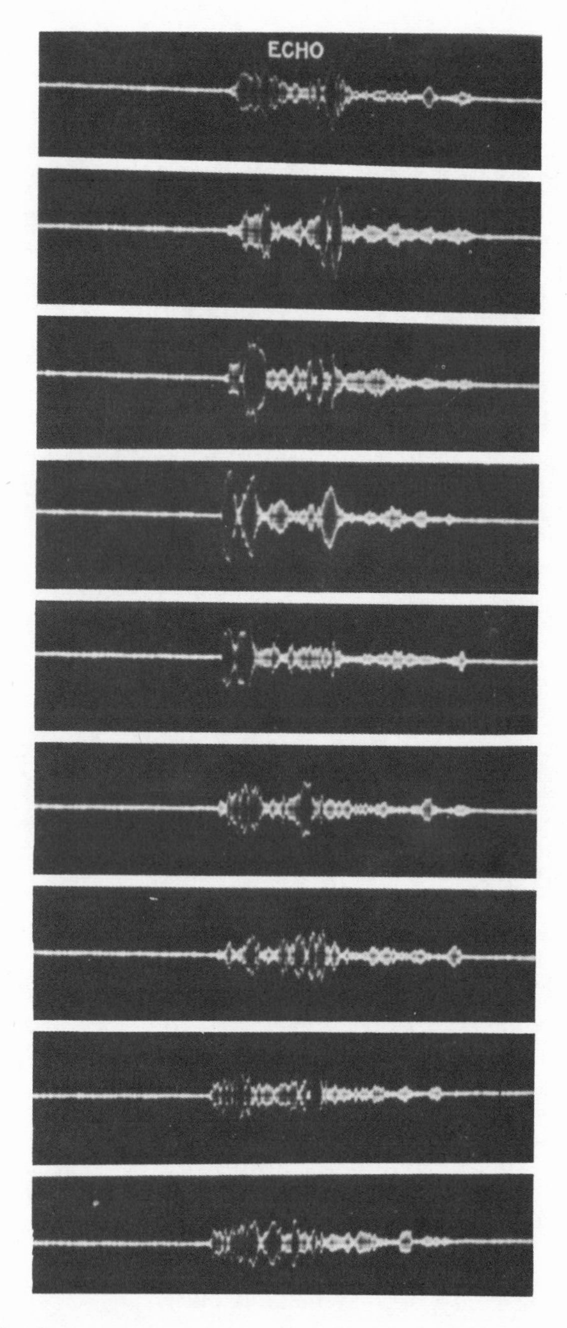

Fig. III-5. Nine consecutive sonar echoes from the same (stationary) target. The differences are attributable to changes in the sound propagation in sea water.

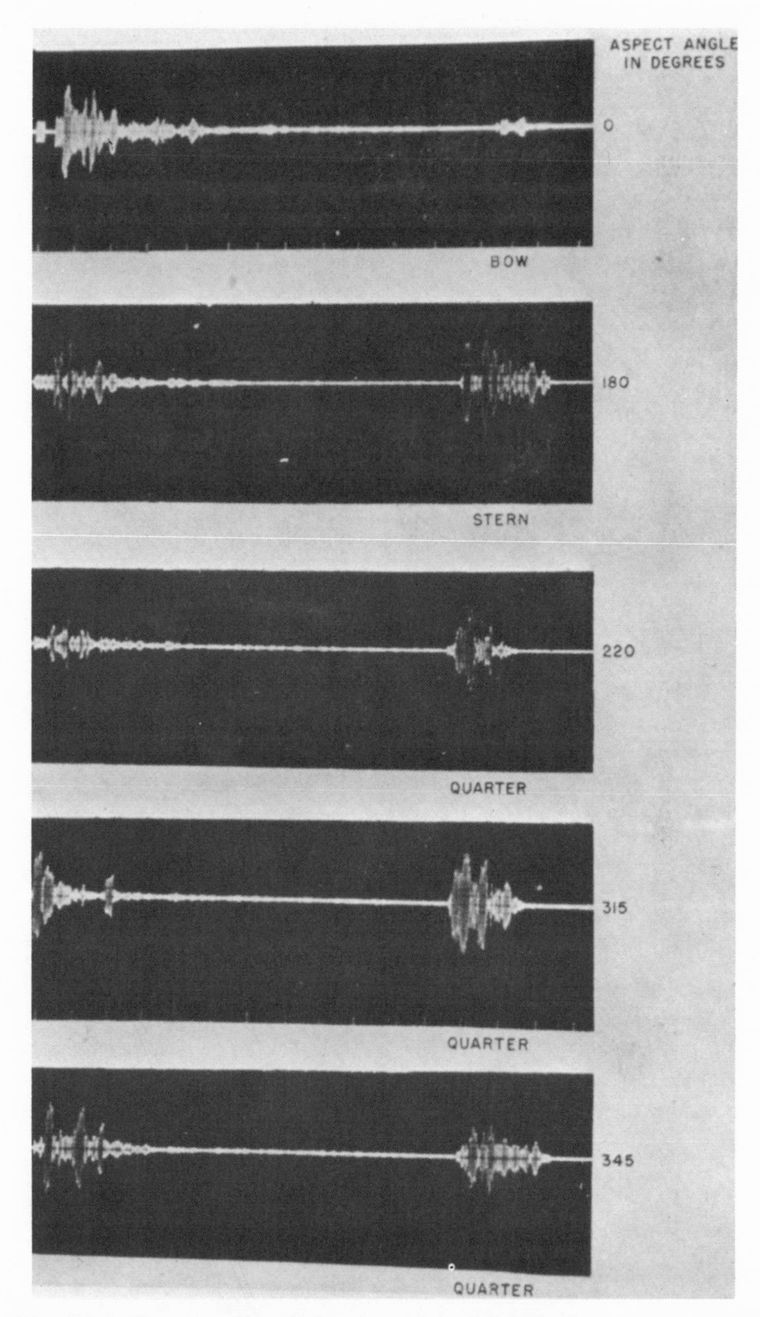

Fig. III-6. The strength and character of a sonar echo from a submarine is dependent upon the aspect angle.

The exact shape of the received pulse echoes often vary with time, depending upon changes in the properties of the water between the sonar and the target (for example, the thermal properties). Figure III-5 shows a set of nine consecutive echoes from the same stationary target. Some similarity can be observed, but marked changes are also evident. The frequency of this sonar signal was 60 kHz and the pulse length was 5 msec. The target was a submarine.

Both the shape *and* the intensity of a pulse reflected from a submarine are dependent upon the aspect of the target; thus, if the submarine target is moving toward the sonar, it presents its "bow" aspect to the sonar and the echo is minimal. This is shown in Fig. III-6. Also shown in that figure are echoes from the other direc-

Fig. III-7. A piezoelectric sonar transducer, used for both radiating and receiving the sonar sound waves.

tions. The stern echo (180°) is seen to be stronger, as are also those reflected off the quarter, shown in the lower three figures.

Directional Properties of Radiators

We saw in Chapter II how radar transmitting and receiving devices often employed large antennas to achieve sharp radar beams and improve the detection-range capabilities of the radar. A similar procedure is employed in sonars.

Figure III-7 shows the piezoelectric "transducer" of a typical sonar transmitter. Piezoelectric refers to the fact that the crystal changes its dimension when an electric voltage is applied across it. A high-frequency voltage is applied to the crystals of this transducer, thus causing it to swell and shrink rapidly at a rate corresponding to the frequency of this impressed voltage. When submerged in sea water, therefore, the expansion and contraction imparts motion to the water immediately in front of the transducer and the sound wave thereby created propagates out as a short pulse of sound.

These crystals also possess the property that when they are stressed (for example, place under compression), they generate a voltage across their faces. The returning echo striking the face thus transmits the forces existing within the sound-wave pulse packet to the crystal transducer. The same electrical connections that were used to apply the alternating voltage during the transmitting situation are thereby provided with an electrical voltage that can be used to indicate the existence of a received pulse to the operator. Because the diameter of the transmitting receiver unit of Fig. III-7 is large compared to the wavelength of the sound transmitted by it, it generates a sharp beam pattern, comparable to the sharp beam patterns of the microwave antennas discussed in Chapter 2.

Figure III-8 portrays a typical beam pattern of such an acoustic transducer (theoretical). In the figure, the expression "DB" is shown. This term corresponds to the widely used "decibel" term which permits the values to be expressed in a logarithmic way. Thus, at the 20-DB point in Fig. III-8, the power radiated along that angle would be 100 times less than the amount transmitted directly on the "beam." Also, as in optics and microwaves, "minor lobes" are

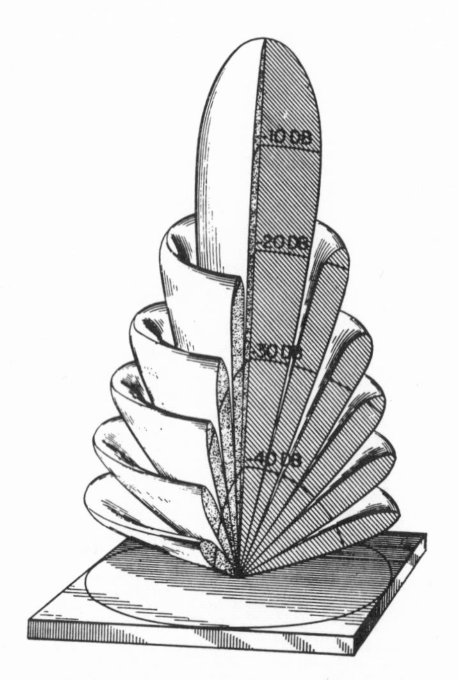

Fig. III-8. A beam pattern of a sonar transducer.

always present. These are shown in Fig. III-8 as the lower conical structures that surround the main "pencil beam."

Passive Systems

In the sonar case, *passive* systems are essentially sensitive receivers for whatever sounds may be present. Thus a submarine uses an array of hydrophones to listen to the destroyer's propeller sounds. Passive sonar, sensitive to much longer ranges, normally provides information on bearing and bearing rate. It can, however, distinguish between many targets when there are several vessels in the area at the same time. Passive devices also have the sometimes critically important characteristic that they obtain information without making telltale sounds or giving away the presence of the receiving vessel. The most widely used submarine sonars are passive systems.

One form of sound that many ships generate orginates at their propellers. In most ships, including submarines, all of the motive power for propelling the ship forward is transmitted through the propellers. A strong interaction, therefore, invariably exists between

Fig. III-9. A spiral pattern of cavitation "bubbles" generated by a moving, rotating, propellor.

the propellers and the sea water. In some instances, the speed of the tips of the blades of the propeller is so high that a phenomenon called "cavitation" results. The phenomenon receives the name from the fact that the forces on the water cause small spherical "cavities" to develop. They are very tiny, resembling small bubbles. However, when the force of the propeller is removed (moving away from that point in the water), these cavities collapse, forming their original "solid" water structure. It is the collapse of these cavities which creates a rather loud noise that can be heard at great distances. Figure III-9 shows a moving propeller creating a spiral set of cavitation "bubbles"; Fig. III-10 shows a twin screw ship generating similar cavitation patterns.

Because one of the great advantages of a submarine is its ability to emply stealth in its operations, such undersea craft employ passive procedures more often than they do active sonar techniques.

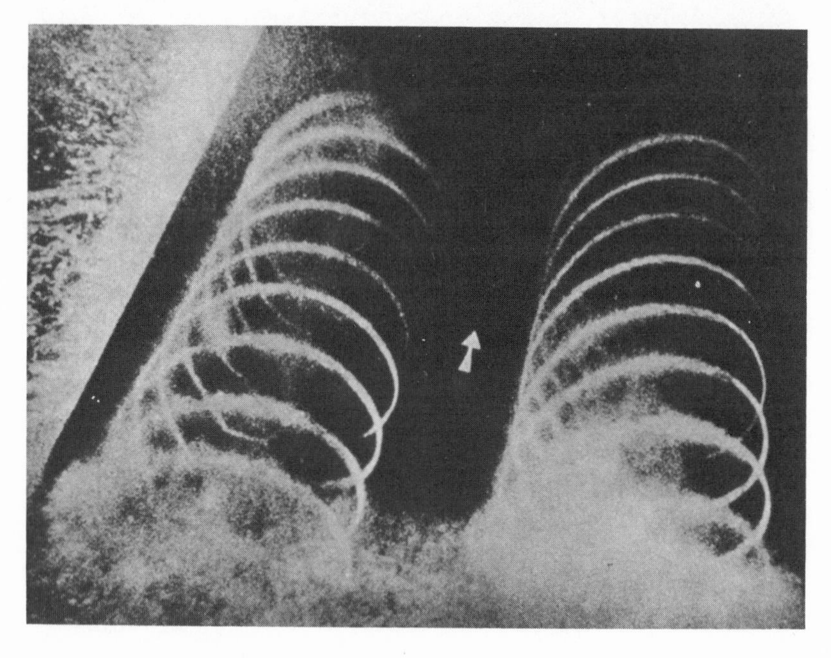

Fig. III-10. Twin propellors create two spiral patterns.

Figure III-11 portrays a "listening" transducer (called a JP unit) designed to provide the bearing relative to the submarine of a noisy target. This is a line array, and it can be rotated about the vertical axis of its support as shown. The operator, using earphones, can detect noisy surface ships with it when its beam is pointed in the

Fig. III-11. A passive, submarine-mounted, sonar array.

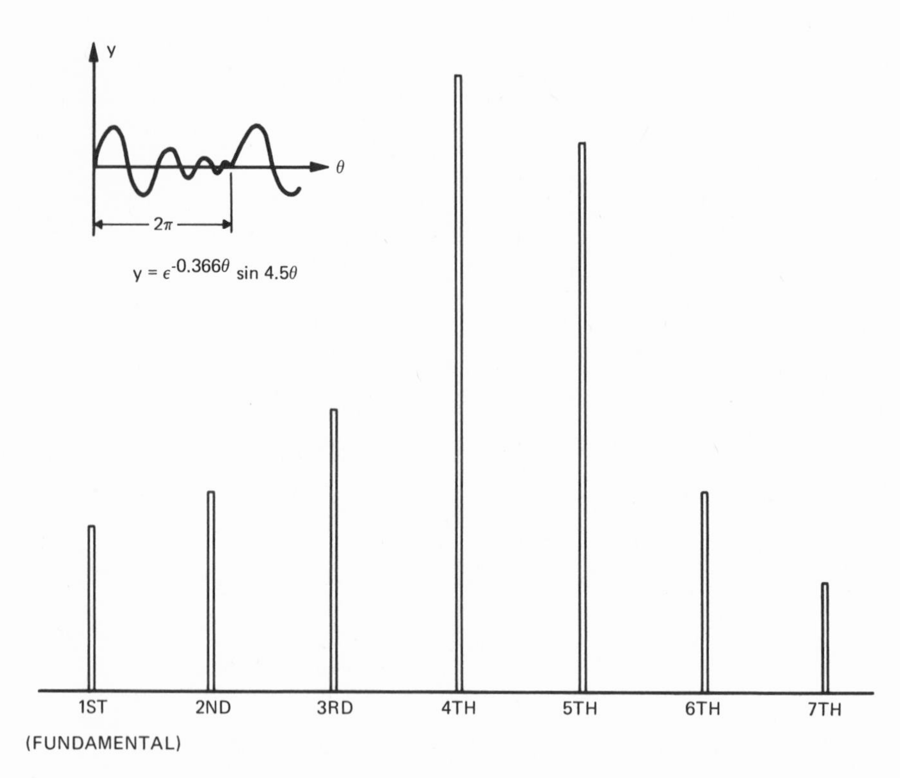

Fig. III-12. The periodic wave (top) can be represented in terms of its frequency components (bottom).

direction of the target. Very often, by determining the rotational speed of the noisy propellers of the surface ship, an estimate of the ship's speed can be made, and by measuring its angular velocity relative to the submarine, some idea of the range of the target can be obtained.

In addition to simply listening by ear, passive sonars often employ a frequency analysis of the sound being received. A frequency analysis of a particular sound is sketched in Fig. III-12. The wave shown at the top is a repetitive one and, therefore, can be analyzed into a series of harmonics. The chart shows the various amplitudes of these harmonics. There are numerous ways of arriving at such frequency analyses. A sharp filter can be made to "scan" the frequency range of interest and its output caused to mark the amplitude of the harmonics. Thus, in Fig. III-12 the filter can be imagined as being moved along the horizontal axis of the sketch. When the position in frequency of the filter matches the position in frequency of the fundamental, the filter would show an output of a certain magnitude. Similarly, as it moved along the horizontal axis, an output would occur each time it reached the various harmonics shown in the figure, with the fourth harmonic generating the strongest output in the filter.

Recently, a technique in which digital computers are used to provide the spectrum analysis has been evolved. Here, as shown in Fig. III-13, the wave is sampled and the magnitudes at the sample points are used to obtain an estimate of the Fourier analysis of the received waves. This process is then called a "discrete" Fourier analysis, and because of the stepwise nature of the digital computers, it fits modern-day computer techniques very nicely.

Figure III-14 shows a plot of frequency versus time of a sound as analyzed by a discrete Fourier process known as the "fast Fourier transform," or FFT for short. Such a presentation permits a sonar operator to observe how the sound patterns being received are changing with time.

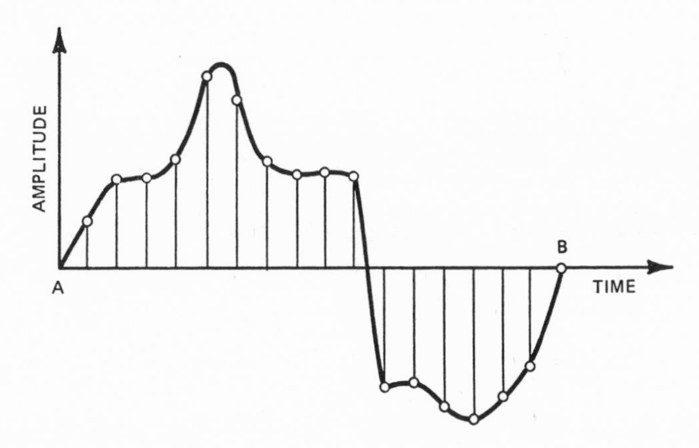

Fig. III-13. The amplitude values of one period of a periodic wave, when sampled as shown, can be used to provide a discrete Fourier analysis.

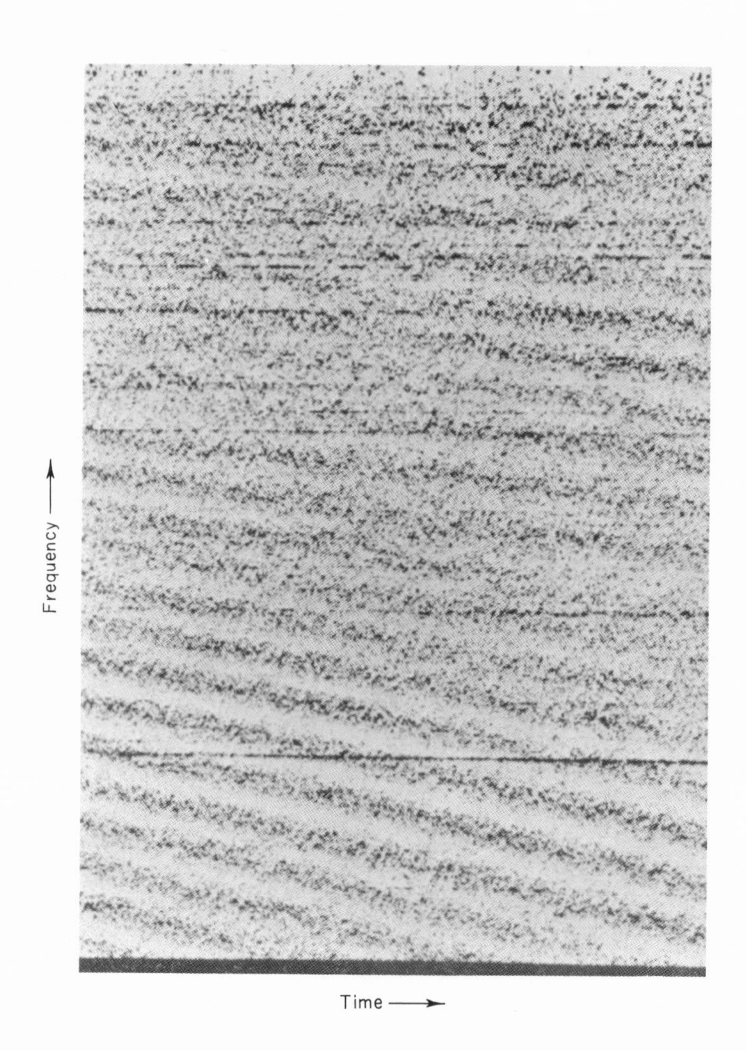

Fig. III-14. This frequency versus time plot of a varying underwater sound was achieved by the fast Fourier transform process (Bell Telephone Laboratories photo).

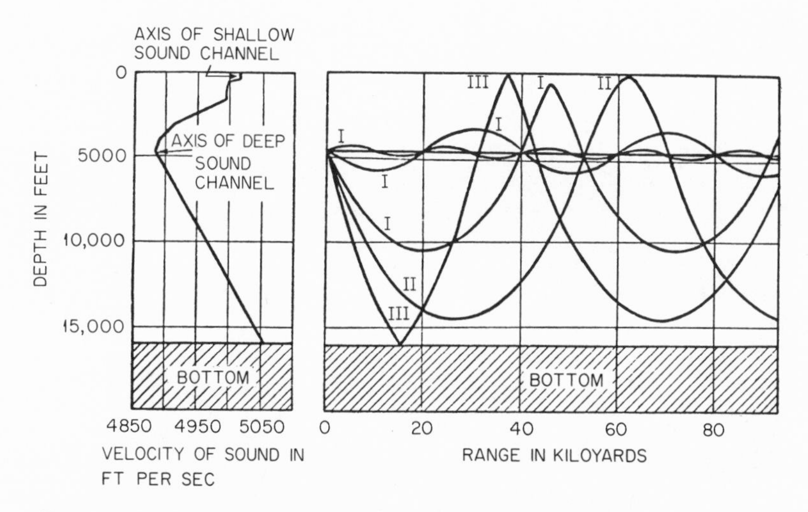

Fig. III-15. The sound-velocity profile in the ocean (shown at the left) causes sound rays to be "channeled" (right).

Sonar Transmission Conditions

Because the water temperature often varies with depth, sonar-receiving efficiencies are quite subject to effects of nature. Figure III-15 shows how this temperature effect causes the sound velocity to vary in the deep oceans. At the left is a plot of sound velocity versus depth. From the surface down to a depth of approximately 5000 ft, the temperature has a negative gradient that causes the velocity to decrease as one approaches this 5000-ft depth. Below that point, the temperature does not change appreciably, but, because the depth is increasing, the weight of the water above is also increasing, and the sound velocity accordingly begins to *increase* again. At the right, various rays (sound patterns) show a concentrated effect of the sound in this "sound axis."

Of more concern to shipboard sonar operations, however, is the situation near the surface. Here, because of the temperature change with depth, sound rays originating at a point as shown in Fig. III-16 are bent downward, thereby creating a region (shown crosshatched) where the sound from that point does not penetrate. A listening sonar located in this crosshatched region would, therefore, not be in a position to hear the sound source shown.

Passive radars are usually employed for the detection and analysis of radio signals coming from stellar bodies, a field referred to as radio astronomy. In wartime, passive radio devices are often used to detect and analyze enemy radar transmissions.

Radar Displays

The function of a radar set is to obtain information about its surroundings such as the distance or range to a given object, the angular bearing or direction of the object (the target), and, often, the angular elevation and the velocity of the target.

The radar information can be fed directly into a computer, recorded on tape or film for later use or analysis, or displayed immediately to a radar observer. The instrument used for this last purpose is the radar indicator discussed above, a cathode-ray tube. The electronic traces on this tube, together with suitable overlays and scale markings, constitute the radar display.

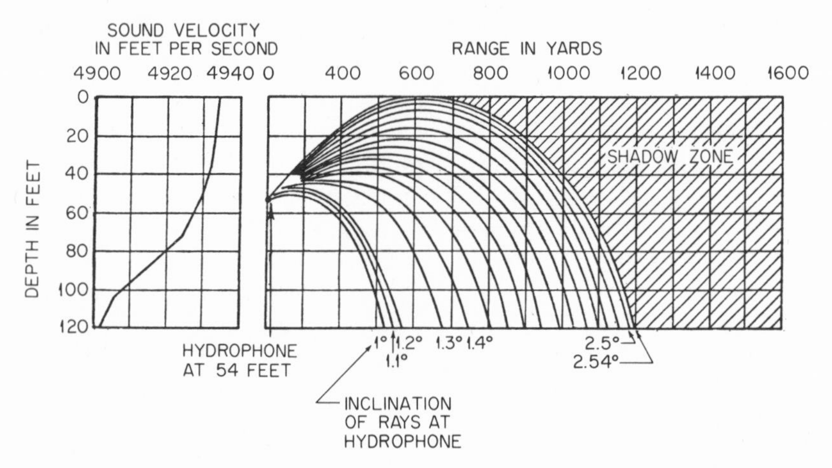

Fig. III-16. Certain sound-velocity gradients that occur near the surface cause sound rays to avoid the "shadow" region shown at the right.

NEW YORK WORLD-TELEGRAM, FRIDAY, JANUARY 25, 1946.

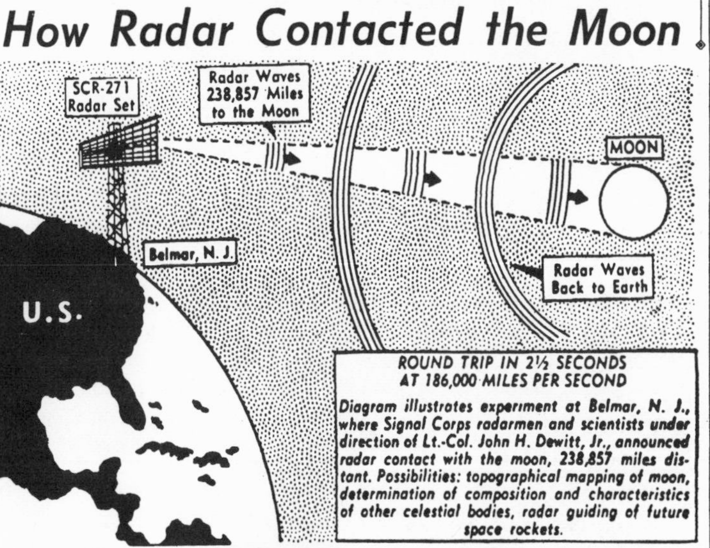

How Radar Contacted the Moon

ROUND TRIP IN 2½ SECONDS AT 186,000 MILES PER SECOND

Diagram illustrates experiment at Belmar, N. J., where Signal Corps radarmen and scientists under direction of Lt.-Col. John H. Dewitt, Jr., announced radar contact with the moon, 238,857 miles distant. Possibilities: topographical mapping of moon, determination of composition and characteristics of other celestial bodies, radar guiding of future space rockets.

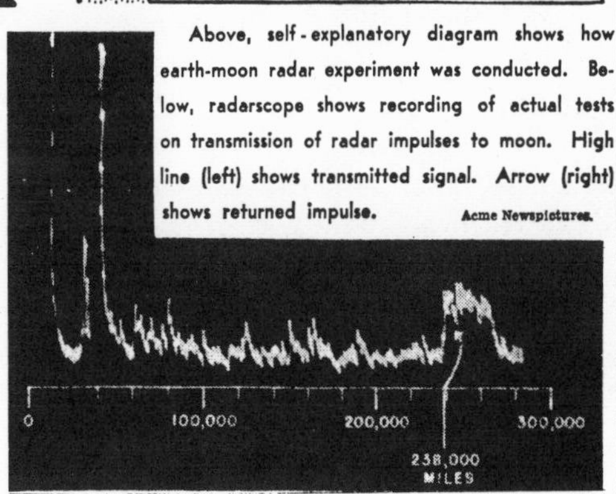

Above, self-explanatory diagram shows how earth-moon radar experiment was conducted. Below, radarscope shows recording of actual tests on transmission of radar impulses to moon. High line (left) shows transmitted signal. Arrow (right) shows returned impulse. Acme Newspictures.

Fig. III-17. A newspaper clipping announcing the first contact of the moon by radar. The A-scope presentation is shown below.

Historically the oldest radar display (called the A-scope) indicates target range and, within limits, the amplitude of the target echo. The A-scope, which is simply an oscillographic presentation of the radar return, is used when angular information is not required or is provided from another sensor. For example, the A-scope might show range to targets directly in front of a radar-equipped aircraft, or to targets whose angular position is determined for the radar with an optical sight or infrared sensor.

Figure III-17 portrays an A-scope presentation of one of the early milestones in radar history. Occurring in January, 1946, the event marked the first time a radar echo was received from the moon. As the upper sketch shows, the transmitted waves traveled 238,857 miles on their way to the moon, and an equal distance after they were reflected. Traveling at the speed of light, they took only 2½ seconds to cover their long path.

At the left of the A-scope record in the lower part of the figure, the writing beam is still being influenced by the very powerful outgoing pulse. After the pulse is on its way, however (indicated in the upper sketch by the sets of 4 lines in the transmitting beam), the writing record shows, through its nearness to the lower scale of miles, that no echoing objects are in the beam. The echo from the moon is the broad plateau at the right. In between the outgoing and echo pulse the writing beam record is quite irregular. This is caused by noise, coming possibly from outer space, but also from circuits within the radar itself. This moon experiment was under the direction of the U. S. scientist, Lt. Col. John H. Dewitt, Jr. The newspaper article describing the experiment also hinted at "radar space ships," which are now a reality. A few months earlier, the British scientist, Arthur C. Clarke, published an article in the October 1945 *Wireless World* entitled: "Extra terrestrial relays: Can rocket stations give world-wide radio coverage?" thus predicting the communication satellites now in operation in many parts of the world.

Another form of radar or sonar display is the one described above in connection with the marine radar (Fig. III-2). Called a plan-position indicator (PPI), it displays bearings to the targets as angles and ranges as the radial distance to the target echoes. For ground installations the PPI angular reference is fixed and usually

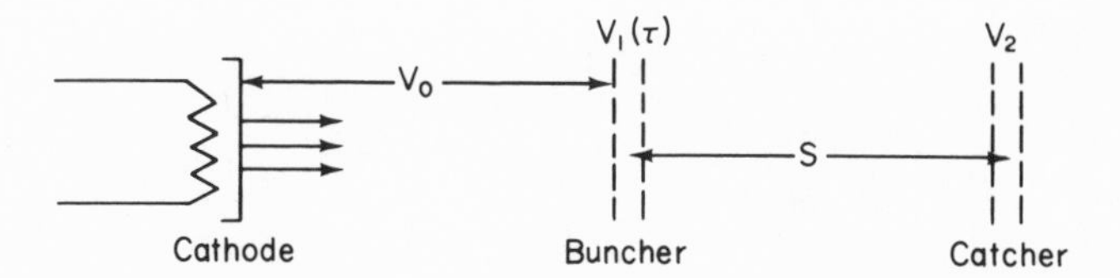

Fig. III-18. By modulating the velocity of electrons emitted by the cathode at the left, many electrons can be made to arrive at the catcher at the right at approximately the same instant.

north oriented. Airborne and shipborne PPI displays can either be north oriented or vehicle oriented, depending upon the application.

Both A-scope and PPI displays are adaptable to sonar systems also.

The Chirp Concept

We observed earlier in connection with Fig. III-4 that a long signal or long pulse generally gave a more distinguishable record whereas a shorter one, although it provided a more accurate range indication for the target, could often be lost in the noise. A form of sonar or radar pulse technique called *chirp* permits the employing of a long pulse without sacrificing the good range capability of a short one.

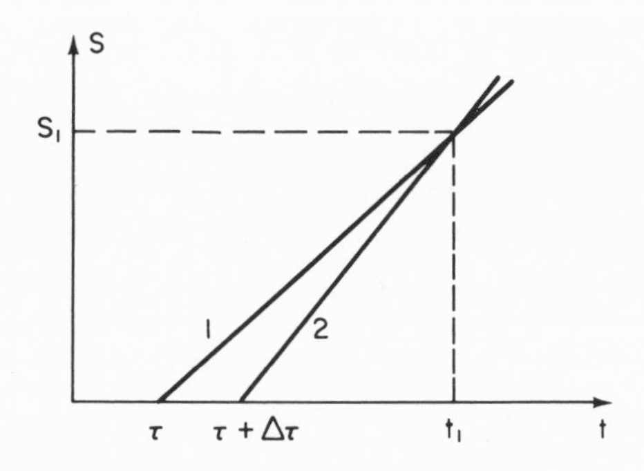

Fig. III-19. Electrons leaving at an earlier point in time can be overtaken by later ones having a higher velocity.

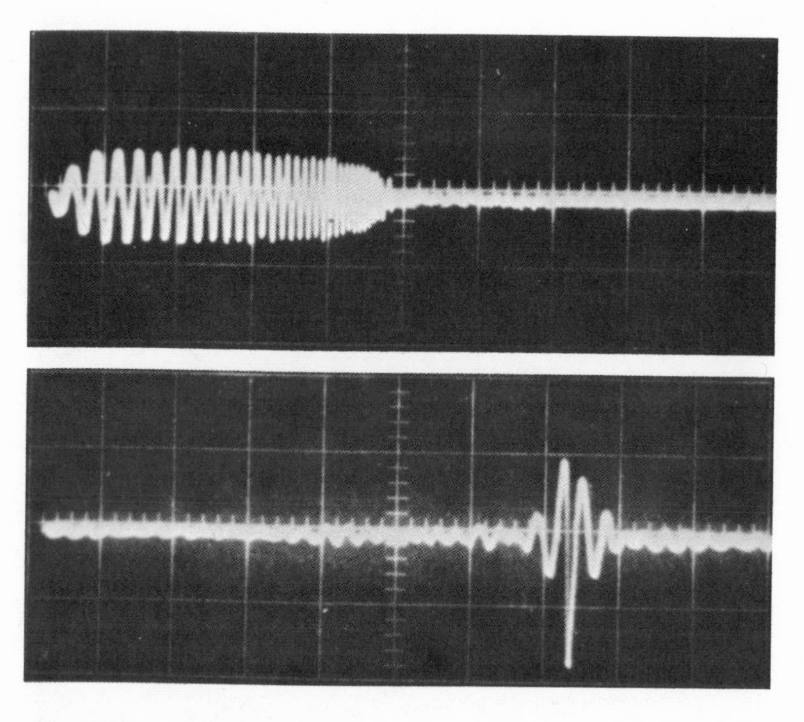

Fig. III-20. A long low-amplitude pulse, in which the wave frequency is made to change during the duration of the pulse, can be transformed into a much shorter, much higher amplitude, pulse.

The concept can be looked upon as having originated before World War II. It was then applied to electron tubes and the process was called "electron bunching." Figure III-18 shows the structure of the electron tube used in this process. A beam of electrons originating in the cathode at the left of the figure moves to the right with velocity V_0 and passes through to structures or grids that constitute the "buncher." These impart different speeds to electrons as they leave. The speed control or "velocity modulation" is accomplished by altering the voltage $V_1(t)$ across the two grids. If this voltage is made to vary in such a way as to cause electrons that leave the buncher at a later time to acquire a higher velocity, they overtake, at the "catcher," electrons that left at an earlier time. Figure III-19 shows how this velocity variation can cause two groups to arrive simultaneously at the point S_1. There the bunched electrons are able to excite a microwave cavity (the catcher) and thus generate extremely high-frequency microwaves.

In the late 1940s, methods for bunching *waves* began to be of interest. In a microwave waveguide, the energy velocity within the guide depends upon the frequency of the microwaves. By varying this frequency, a velocity modulation effect can again be achieved whereby waves fed into a waveguide at a later time can be made to catch up with earlier, lower-frequency, waves. Bell Laboratories scientist Sid Darlington recognized that this phenomenon would permit an outgoing sonar or radar pulse to be made quite long, but by endowing it with a varying frequency over its length, it could be manipulated, *when received as a target echo,* so as to become quite short. The frequency variation needed was one in which the frequency or pitch rose during the lifetime of the pulse. Since this is not unlike the rising chirp of certain bird calls, the name "chirp" has been adopted as a description of this way of converting (after reception) a long pulse into a short one. Figure III-20 shows how effective such a pulse-compression procedure can be.

Chapter IV

TYPICAL RADAR AND SONAR SYSTEMS

Harbor Radars

In the previous chapter, a marine radar using a plan-position indicator display was described. A similar radar could be used as a shore-based system for harbors, thus informing the harbor master of the location of approaching and departing ships. One advantage of radar results from the fact that most radio waves are not blocked by fog or heavy rain, as light is, so that a radar can "see" objects even when they are completely hidden in dense fog.

Aircraft Radars

Plan-position indicator radars are also often used in the control and surveillance of air traffic in civil applications and in the detection of aircraft and missiles in defense applications. In the civil applications special radars are often employed to enable aircraft approaches and landings to be carried out under the direction of a ground-based radar operator.

By choosing radio waves having very short wavelengths, radar echoes of rain clouds can be obtained. Such radars can thus detect the presence, location, and (with certain qualifications) intensity of precipitation. Ground installations make possible the detection and tracking of storms, and airborne weather radars alert pilots to the presence of severe weather in and near their path. These radars are usually called "weather radars."

Fire-Control Radars

In Fig. I-11 an 8-ft parabola was shown. That photo was taken on May 25, 1943, when the first complete systems test of a fairly important World War II radar was being conducted at the Radar Research Laboratory of the Bell Telephone Laboratories at Holm-

Fig. IV-1. A photo of the experimental antenna of Fig. I-11 as mounted for actual tests at sea.

del, New Jersey. This radar, an experimental model of the Mark 13, was one of the early fire-control radars, destined to be used to direct the main batteries of battleships and cruisers. Another view of this experimental version is shown in Fig. IV-1. The tubular structure is seen to have attached to it a horn feed, and in operation the entire 8-ft reflector and its feed horn were made to oscillate in the horizontal plane with a scanning motion of $\pm 5.8°$ at an oscillation frequency of 5 Hz. The choice of these parameters was dictated by the radar mission; it was to display a small sector (11.6° wide in azimuth, as contrasted with the full 360° of Fig. III-2), but it was to display the information in that sector rapidly (5 back-and-forth scans, or 10 total scans, per second). This enabled the radar operation (and the fire-control officer) to "see" in both range and azimuth, the plumes of water that arose as the main-gun battery shells struck the water. Since the radar also "saw" the enemy ship under attack, the fire-control officer could then adjust his gun batteries so that the next salvo of shells would zero in on the ship under attack. All of this information was available during the black of night, during fog, etc. Because of its great effectiveness to the Navy, this radar was highly commended by Rear Admiral G. F. Hussey, Jr., then Chief of the Bureau of Ordnance. In a letter dated June 25, 1945, he stated: "While fleet experience with the radar equipment Mark 13 is limited to only a few ships, reports from these ships indicate that it is considered the best radar equipment yet installed on shipboard."

The advent of radar proved to be extremely important in Naval applications. For example, in the autumn of 1942, the U.S.S. Boise sank six opposing war ships in a night battle through its use of radar. The battle took place on October 11 off Savo Island near Guadalcanal. Even in the intense darkness, the Boise was able to contact the enemy with its surface-search radar equipment, an early type developed at the Bell Telephone Laboratories and manufactured by the Western Electric Company. As explained by Lieutenant Commander Philip C. Kelsey, "We contacted the ships at 13,500 yards, according to our surface-search radar, and thus provided the fire-control party with initial target bearings. The enemy ships, apparently unknowing in the darkness, moved closer and closer to the Boise until a scant 3900 yards separated us." The Boise opened fire at almost point blank range and, in 27 short minutes, the engagment was over with none of the enemy ships remaining afloat.

Early in the spring of 1942, the Bell Laboratories had set up a test facility at Atlantic Highlands, New Jersey, overlooking Sandy Hook Bay and the entire New York Harbor, for the purpose of testing radars against Naval vessels. According to the Bell Laboratories Record, these facilities played a great part in the convincing of high ranking Naval officers, particularly those to whom radar was an unknown quantity in the early days of the war, of the fact that radar could accurately direct the pointing of ships' guns. The actual demonstration of fire-control radar was found to be a much more satisfactory method of instruction than words, charts, and pictures could be. Figure IV-2 is a photograph of this facility at Atlantic Highlands. At the left of the figure was one of the first Naval fire-control radars. Immediately to its right is the Mark 13 radar, which was just discussed, in its production form.

Fig. IV-2. An early radar testing site.

Fig. IV-3. The production model of the experimental radar shown in Fig. I-11.

Fig. IV-4. The radar of Fig. IV-3 with its weather cover (radome).

Figure IV-3 is a close up of this production model and Fig. IV-4 shows the same unit with the plastic radar cover (called a radome).

Figure IV-5 indicates how this radar was mounted on shipboard. The equipment used by the radar operator is shown in Fig. IV-6.

Other Forms of Shipboard Radar

Figure IV-7 is a photo of the SJ submarine radar antenna. This enabled a submarine to remain almost completely submerged and to employ the radar in the dark of night when visual observation by means of a periscope was not favorable.

Fig. IV-5. The radar of Fig. IV-3 mounted on shipboard.

Fig. IV-6. The electronic equipment associated with the radar of Fig. IV-3.

Modern Naval vessels carry innumberable kinds of radars for detecting oncoming aircraft and other surface ships.

Space and Astronomy Radars

Radar has played a vital role in tracking earth satellites and in making important measurements of their orbit paths. It is also making contributions in astronomy; radar returns have been obtained from the moon, Venus, and the sun. Radar has made it possible to make studies of radio propagation in space and to investigate such features as the surface roughness of the moon and its vibration motion.

Sonars

Sonar systems are widely used by the U.S. Navy and other navies, appropriately designed equipment being installed in destroyers, escort vessels, submarines, cruisers, and fast merchant vessels. Sonar systems have been designed for operation from helicopters, seaplanes, and lighter-than-air craft, and even for use by amphibious forces and by frogmen.

Figure IV-8 is a photo of a sonar called the QGB sonar. It has magnetostrictive transducers operating in the frequency range from 17 to 26 kHz, permitting its use as an active sonar, a listening son-

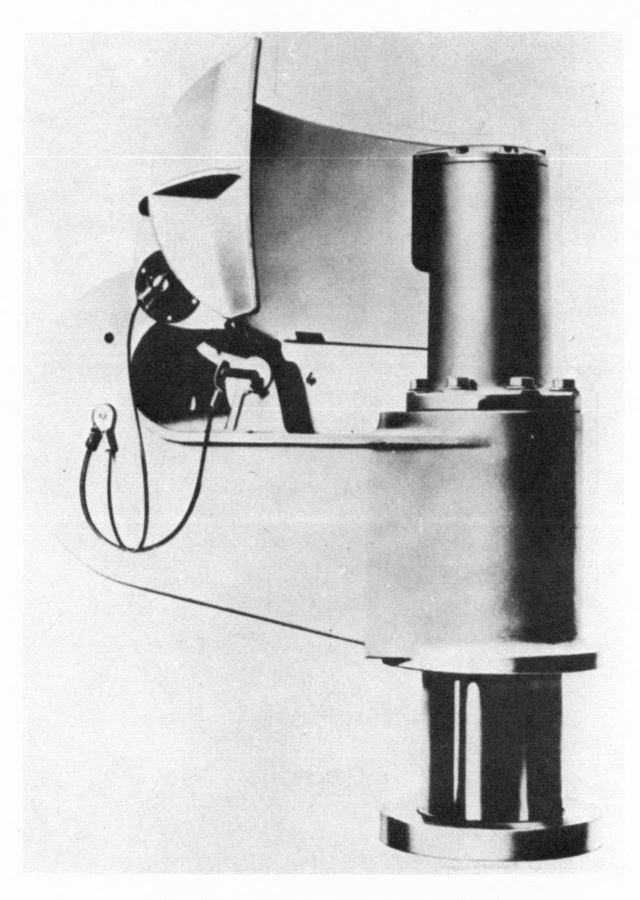

Fig. IV-7. A submarine-mounted radar.

ar, or an underwater communications facility (to a friendly submarine or other surface ship).

Figure IV-9 portrays a ship utilizing a towed array. By being

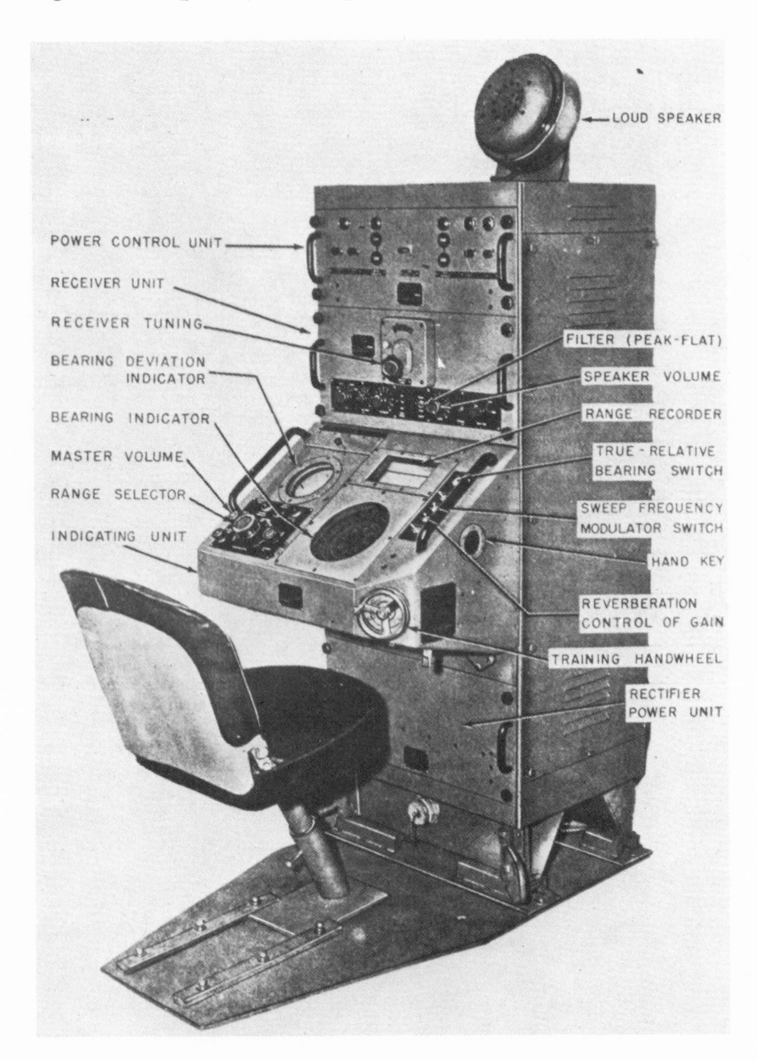

Fig. IV-8. The electronic equipment associated with the QGB sonar.

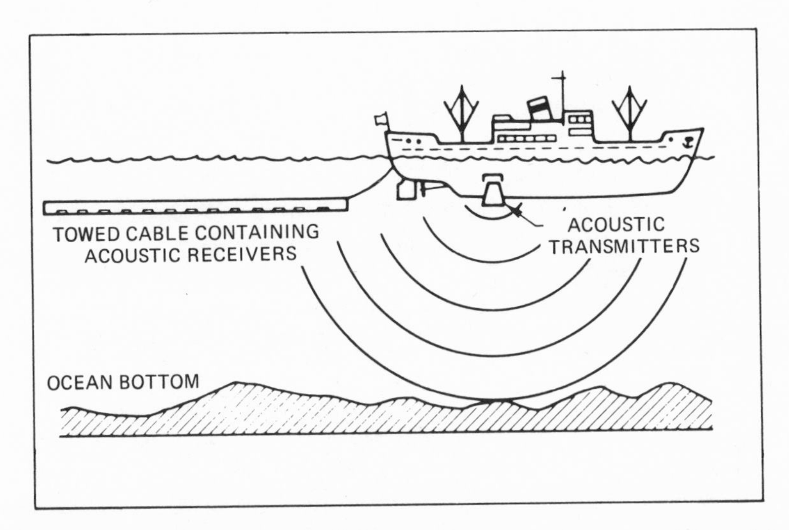

Fig. IV-9. A towed array (left) can be used as a passive (listening-only) sonar, or as an active one, in combination with a transmitting transducer (right).

towed in a straight line behind the ship, it can generate very sharp beams and can be used either for listening only (passive use), or, as indicated, in conjunction with an acoustic transmitter on the ship.

During World War II, the type JP sonar described in the previous chapter was developed. As noted, it was a listening sonar employing a mechanically trained line hydrophone with a selective amplifier leading to the operator's headphones. Since World War II, development of submarine sonar equipment has proceeded along the lines of physically larger arrays, better matching of hydrophones to improve the beam patterns, and more sophisticated information processing in the receivers.

An important improvement was made with the development of scanning sonar. In this system the sound pulse is sent out in all directions while the receiving beam is rapidly rotated to give a spiral scan presentation. Returning echoes are displayed on the screen of a plan-position indicator scope, and give a picture of underwater targets closely analogous to that given on the radar screen.

Radars for Air Traffic Control

Radar has proved very valuable in air transport applications. In 1967, the air traffic control system operated by the Federal Aviation Agency in the United States included 327 control towers, 127 terminal radar control facilities, and 333 flight service stations. These facilities were manned by more than 17,000 operating and support personnel.

Of the 127 terminal radar control facilities, 91 were directly associated with specific control towers, and the remaining 36 were separate Radar Approach Control and Radar Air Traffic Control Center units at Air Force or Navy bases. There were also 89 long-range air route *surveillance* radars, many remotely located, and connected to the appropriate center by microwave links. In both terminal and en-route systems, radar data are displayed either on low-brillance plan-position indicators or on newer, brighter display systems.

Landing Aids

On-board radar systems can be used to permit aircraft to achieve an all-weather landing capability; however, from the viewpoint of flight safety it is better to have ground-based cooperation for this critical phase of flight. Low-visibility, automatically controlled approaches and landings are technically feasible, but the main impediment to the deployment of such automatic landing systems has been the assurance of equipment reliability. Approach guidance provided by the present ground-based radar systems is satisfactory for most automatic landing requirements.

Guidance schemes that can be considered as alternates to the present systems should provide the equivalent of a fixed path for the aircraft to follow. Ideally, a ground-based, precision-approach radar should assist, by monitoring, any aircraft employing an automatic approach so that the measured aircraft position can be compared with the most satisfactory path. It appears that new concepts in pilot training may be needed for the era when on-board automatic landing equipment will play a major role in aircraft flight control.

Collision Avoidance

Another area of radar technology that is directed at a major national problem is collision avoidance. Current developments are only concerned with protecting commercial transport aircraft from each other. Such a partial solution is certainly useful since a reduction in the frequency of this type of accident would result in huge savings through reductions in equipment loss and settlements.

Chapter V

DOPPLER AND PHASED-ARRAY SYSTEMS

The Doppler Effect

For sound waves, the Doppler effect can be observed as a drop in the frequency or pitch of the whistle of a rapidly moving locomotive as it passes the listener. While the sound source, the locomotive whistle, is approaching, the sound possesses an up-Doppler, that is, an increase over the pitch that would have been observed had the relative motion between sound source and listener been zero. After the locomotive has passed the listener and the source is receding, a down-Doppler is observed, resulting in the listener perceiving an apparent drop in the pitch of the whistle.

Radio energy reflected by a moving target that has a component of motion toward or away from the radar illuminating the target similarly exhibits, because of the Doppler effect, an apparent change in frequency on reception. This phenomenon can be expressed analytically: the shift in frequency f_D is proportional to the

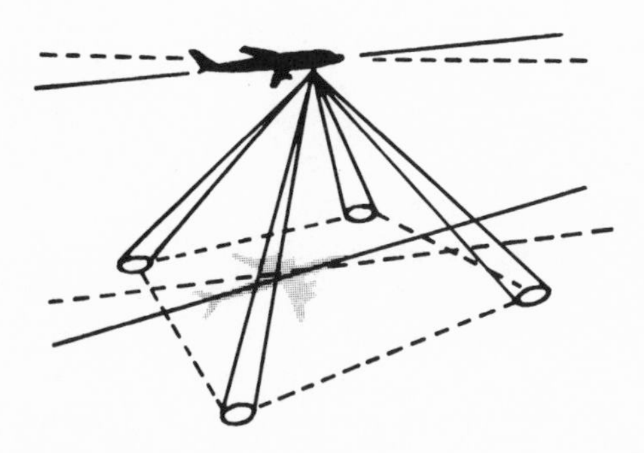

Fig. V-1. The Doppler radar of this aircraft has four radar beams directed towards earth. The plane's motion results in the return signals having their frequency changed in ways that inform the pilot of the presence of head winds, crosswinds, or tail winds.

relative radial velocity v and the original frequency f_0, that is, $f_D = 2vf_0/c$.

In radar employing very short pulses, such variations in frequency are usually too small to be significant. However, in special applications, using equipment designed to distinguish small differences in frequency between the transmitted signal and its echo, advantage can be taken of this effect to determine relative radial velocity. A particularly valuable use of this phenomenon is the aircraft navigation technique commonly called "Doppler radar" or "Doppler navigation."

In Doppler navigation systems, four very narrow beams of microwave radar energy are radiated from the aircraft as shown in Fig. V-1. The frequencies of the four reflected signals are continuously compared with the frequency of the signal originally transmitted, and the amount of Doppler in each beam is calculated (usually by an electronic computer). If there is no crosswind, the direction of heading of the aircraft correspond to its course direction, and the echoes received in the two forward beams will experience an equal amount of Doppler (an up-Doppler) which, for level flight, specifies the plane's forward velocity relative to the earth. This may be quite different from the plane's *air* speed because of the existence of a head wind or tail wind. Thus, the speed relative to the air (the air speed), for a strong head wind, might be quite large, with the effective speed relative to the ground being much less.

The two rear beams will also observe a similar Doppler, theirs, however, being a down-Doppler, and, again for level flight, this down-Doppler will again specify the plane's ground speed. If the plane is not on level flight, these two will differ, permitting the pilot to be alerted to this fact, but, in addition, the plane's correct ground speed can still be ascertained from these two Dopplers.

When a crosswind exists, if the pilot follows his compass and continually flies, say, due east, he will find the crosswind pushing him off course to one side or the other even though his compass says his path is correct. His Doppler radar will however, alert him that he is being blown sideways, because the two forward beams will show different up-Dopplers. Thus, if the plane is being blown off to the left of course, the left forward beam will show a *higher* up-Doppler than the right one. (They will both, of course, still exhibit the up-Doppler due to the forward motion). Similarly, for the case assumed, the two rear beams will exhibit different *down-Dopplers,* the left one showing less than the right. The radar thus enables the pilot to fly not only a *level* course, but, by its specifying the amount the pilot should alter the direction of the plane during a crosswind (that is, by "crabbing") and thus cause the four observed Doppler effects to be as needed, he can accurately navigate regardless of head winds, tail winds, or crosswinds.

Doppler Navigation Equipment

Antennas for Doppler navigation systems are usually fixed planar arrays designed for rigid mounting to the aircraft structure and comprising no moving parts. Their thin profile permits mounting them in a semiflush position beneath the aircraft.

The computer equipment consists primarily of a frequency "tracker" for deriving from the Doppler information the ground speed and drift angle. Often more sophistication is added, permitting the computer, in its display, to present the pilot with information on "miles to go" to destination, and miles to the left or right of a selected course.

Other Doppler Applications

Use has been made of the Doppler technique in ground-based radars for distinguishing between moving and fixed targets, and also for determining the speed of targets. Although of particular value in the case of ground radars detecting aircraft targets (because of their high speeds) the procedure can also be used on targets having much slower speeds. One such application is in the determination by radar of the speed of automobiles; such techniques are currently used in numerous cities to control the volume of traffic on expressways and to enforce speed laws.

Fig. V-2. An experimental radar mounted on the front end of an automobile.

Radars for Cars

Whereas the high-power sophisticated radars for military systems are usually very costly, developments in microwave technology have made possible very low-cost (low-power) systems that are now finding greater and greater use in automotive applications. The automotive industry is particularly interested in developing automatic braking systems utilizing radar.

One such system under development utilizes an "adaptive" speed control, controlling both the throttle and brake. Figure V-2 shows a system designed to test the principles involved (installed in the front of a Volkswagen) and Fig. V-3 shows the radar horns of an actual development model installed in a Lincoln car. The driver selects a (constant) speed which the car maintains on the open road. Then, when the car's radar detects a second car in front of it, the radar "adapts" the actual speed of the car so as to maintain a safe distance between it and the car in front. (This safe distance, in feet,

Fig. V-3. Two small radar horns on the front of an automobile provide a means for achieving a safe and automatic speed control (Bendix Corp.).

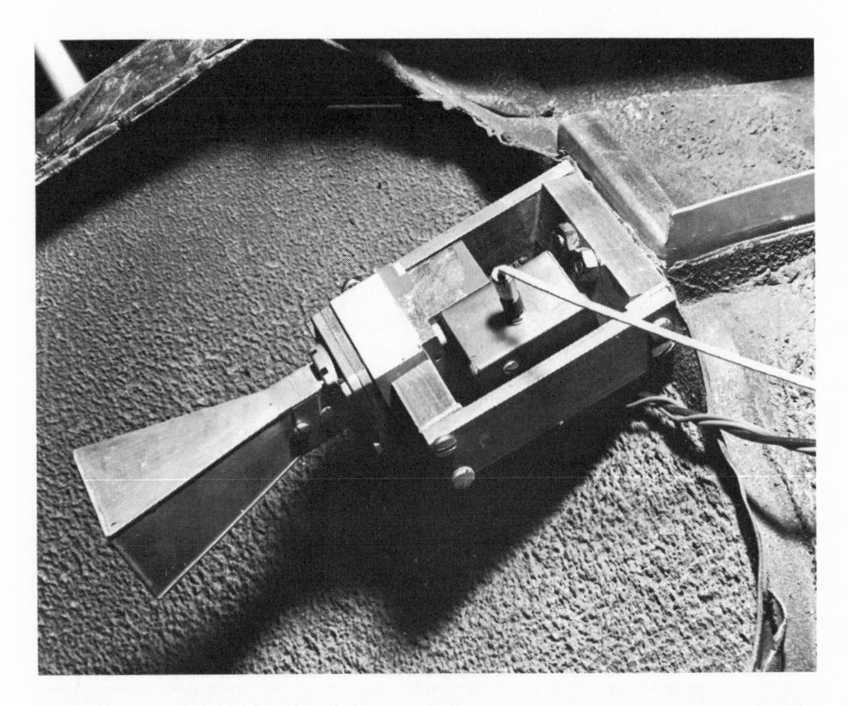

Fig. V-4. A tiny, very inexpensive, radar can measure the rotational velocity of each car wheel, thereby helping to provide antiskid performance (Bendix Corp.).

is usually chosen to be equal to the speed, in miles per hour, plus 30; thus at higher speeds this "safe distance" increases.) The radar measures both the speed and the range of the car in front by means of the Doppler shift, using two sets of transmitted pulses spaced apart slightly in frequency.

The use of small, inexpensive radars is also being explored for "antiskid" applications in cars. This involves a concept first utilized in aircraft landing brakes. On icy roadways or airport runways, applications of braking pressure could cause one wheel and not the other to *skid,* with the result that the stopping force provided by that wheel becomes extremely small. The plane or the car then usually swerves dangerously because of the lack of balanced braking.

The antiskid provision causes the rotational speed of each of the wheels to be ascertained, and when the speed of one wheel begins to drop below that of the others (that is, as it is just beginning to skid), the speed sensor causes the brake pressure on that wheel to be lowered, thus eliminating its tendency to skid.

Figure V-4 shows a tiny radar installed on an automobile for the purpose of accurately measuring the wheel's rotational velocity. Each wheel would be equipped with a similar tiny and inexpensive device for detecting wheel-speed changes on initiation of a skid. The advent of two new inexpensive microwave generating devices, the Gunn diode and the avalanche diode, played a large part in the development of small, low-cost radars of this kind.

Phased-Array Radars

We discussed in Chapter II, in connection with Figs. II-21 and II-22, how the beam of a broadside radiator, comprising many small radiators arranged in an *array,* can be steered by incorporating various amounts of delay in the lines energizing each elemental radiator. We also noted in Chapter I that, for single-frequency waves, a shift in phase of a certain amount is equivalent to a time delay. Thus, if a wave has a wavelength of 4 ft, a delay of 1 ft is equivalent to a phase shift of 360°/4, or 90°. Similarly, a delay of 2 ft is equivalent to a 180° phase shift.

In connection with the stepping of lenses discussed in Chapter II, we saw that a phase shift of 360° (i.e., a one-wavelength step) caused no significant change in the radiated pattern of the lens for waves possessing the design wavelength. For the elemental array under discussion, a similar procedure is possible when the delay or phase along the array is gradually changed. At that point where the change in delay equals one wavelength (360° phase change), the delay can be reduced to zero and the process of increasing the delay again commenced.

A sketch of this procedure is shown in Fig. V-5. Just as the prism of Fig. II-24 causes the sound-wave beam to be tilted downward, the prisms of Fig. V-5 tilt to the right the plane waves arriving at the bottom of the figure. Here, however, the maximum thickness of each prism is made such that the maximum delay it introduces is equal to one wavelength (or 360°). Proceeding from left to right, waves emerging from each prism are successively "ahead," by one, full wavelength, of the waves emerging from the prism to its left. Accordingly, the wave crests and wave troughs remain perfectly aligned (for waves of the design wavelength).

One can also imagine in Fig. V-5, many individual radiators, as in Fig. II-20, placed at the line corresponding to the lower surface of all the prisms, and all of these energized such that the *phase* of each is proper to generate the tilted wave fronts at the top of the figure. If there were, say, 10 radiators lined up where each prism is sketched, each one would be successively phased 36° behind the one on its left, so that radiator number 10, located at the thickest part of the prism, would be shifted by 360°, or one wavelength. When one reaches the 360° point of a circle, one starts over again from 0°, and the procedure would be repeated for the next ten ra-

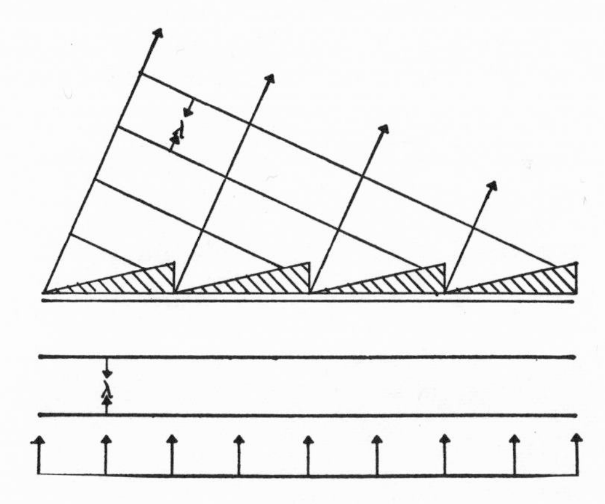

Fig. V-5. Single-frequency waves arriving from the bottom of the figure have their direction of advance altered by the prisms. The maximum thickness of the prisms corresponds to a delay or phase change of one wavelength (360°). Had there been only two prisms, of the same thickness but twice as wide, the change in direction of the beam would have been only half as much. If one imagines the lower arrows to be elemental radiators of a line array, one sees that the imparting of a continuous phase change, from left to right, to these radiators would achieve the same beam tilt as the prisms do.

Fig. V-6. A World War II phased-array radar. The phase of the microwave source signal fed to each of the rod radiators shown could be varied, thereby enabling the horizontal direction of the beam to be varied.

diators, placed where the second prism (from the left) is located in Fig. V-5.

Because there are fairly simple electronic means for providing to the radiators the phase shifts we have just discussed, this technique

Fig. V-7. A photo of one of the rod radiators ("polyrods") of the radar of Fig. V-3.

Fig. V-8. An exposed rear view of the radar of Fig. V-3, disclosing the waveguides in which rotary phase shifters enable the phase of the rod radiators to be varied, thereby shifting the radar-beam direction.

is often referred to as "electronic-beam steering" of phased-array radars (or sonars).

Figure V-6 shows a phased-array radar employed as a main battery fire-control radar during World War II. In this system the radiators are the "polyrods" discussed in Chapter II in connection with Fig. II-23. Figure V-7 is a photo of one of these individual polyrod radiators.

Figure V-8 shows the rear of this radar and the numerous waveguides feeding the individual radiators. Rotary phase shifters are incorporated in these waveguides, permitting the phase of each radiator to be altered in accordance with the beam pointing angle desired, as was discussed in connection with Fig. V-5. Thus, without moving the array, the beam can be pointed in any desired direction within limits. The rotary phase shifters permit a rapid change in the pointing angle to be accomplished, so that the beam can be rapidly scanned in the horizontal direction. Its action is thus similar to the mechanically rocked dish of the Mark 13 radar shown in Fig. I-11 and discussed in Chapter IV.

The huge radar of Fig. II-25 is also a phased-array radar, and its transmitting and receiving beams can similarly be steered electronically. It has the additional ability to alter the pointing of its beam in *both* vertical and horizontal directions. Many sonars are also phased-array systems. Figures V-9 and V-10 show an array of underwater transducers. Through phasing procedures the beam of this radiator could be steered. Phased-array systems possess the capability of generating simultaneously, and with the same set of array elements, a multiplicity of beams, each of which can, if desired, be "scanned," that

Fig. V-9. An underwater acoustic array about to be lowered into the water from the transmitting ship.

is, aimed so as to follow a moving target. This capability is obviously not one possessed by a parabolic dish antenna. Some phased-array passive sonars utilize this ability to form many receiving beams from the same set of array elements.

Fig. V-10. Other views of the sonar array of Fig. V-9

With the rapid growth in solid-state (transistor) electronics, several recent phased-array radars use solid-state transmitting and receiving elements. Some incorporate hundreds of these solid-state devices, so that even though the power output of each individual transistor cannot compare with the higher power of vacuum-tube microwave generators, their *summed* power is sizeable. Also, these radars often use the pulse compression procedures described in Chapter III so as to cause a long, low-amplitude transmitted pulse to be shortened to a high-amplitude one. Signals radiated from such all solid-state phased-array radars at X-band (the frequency region of about ten thousand million hertz) can thus reach *effective* peak powers of tens of kilowatts.

Because the extent to which each element of a phased-array system can be energized, the shape of the beam pattern that the array radiates can be controlled. We saw in Chapter II that an "amplitude taper" incorporated in a radiating device could cause the side lobes to be reduced. The advantage of this result can be seen from the following considerations. In a radar, a very *strong* reflecting target positioned in the direction of a potent side lobe will produce, in the receiver display, a received signal indistinguishable from a signal generated by a weakly reflecting target lying within the main beam (the *major* lobe). Thus, by reducing the minor-lobe intensity, the chances of this result taking place are minimized. In phased-arrays, the control of the element amplitudes permits a very effective reduction in the side-lobe levels.

We see that three distinct advantages of phased-array systems are: inertialess scanning, multiple beam formation simultaneously with the same array, and effective side-lobe level control.

Chapter VI

HOLOGRAPHY

As we have seen, the two technologies of radar and sonar have been patterned along the lines of the classical optical science of photography, with all three technologies employing extensive use of lenses and reflectors. Radars and sonars have in the past accordingly inherited the limitations of photography.

A new form of radar, coherent or synthetic-aperture radar, has broken away from traditional approaches, and has thereby freed itself from many of the limitations of photography. This is because coherent radar is based on the new technology of holography. As the awareness among engineers of the basic *differences* between holography and photography (and similarly between coherent radar and ordinary radar) has grown, new developments in radar and sonar have blossomed.

We shall see that one of the unusual capabilities of holography is its ability to record not only three-dimensional pictures, but also pictures in which all objects, near and far, are in sharp focus. Similarly, just as the optical hologram brings each point of a three-dimensional scene into sharp focus at any distance from the photographic plate, the microwave hologram of a synthetic-aperture radar record provides good focus on all its reconstructed points.

This capability, common to both coherent radar and holography, sets them apart from ordinary radar and photography. The "depth of focus" limitation in photography is traceable to the use of lenses or paraboloidal reflectors for which only one plane section of the image field can be recorded in truly sharp focus, and this limitation does not apply to holography or synthetic-aperture radar. Let us, therefore, examine the hologram process, and see how it has helped radar and sonar find new performance horizons.

Making a Hologram

In forming a *hologram*, two sets of single-wavelength light waves are made to interfere. One set is that issuing from the scene to be photographically recorded; almost invariably, it is an extremely complicated one. The other is usually rather simple, often being a set of plane waves. This second set is called the reference wave, and, in reproducing or *reconstructing* for the viewer the originally recorded scene, a similar set is used to illuminate the developed photographic plate, the hologram.

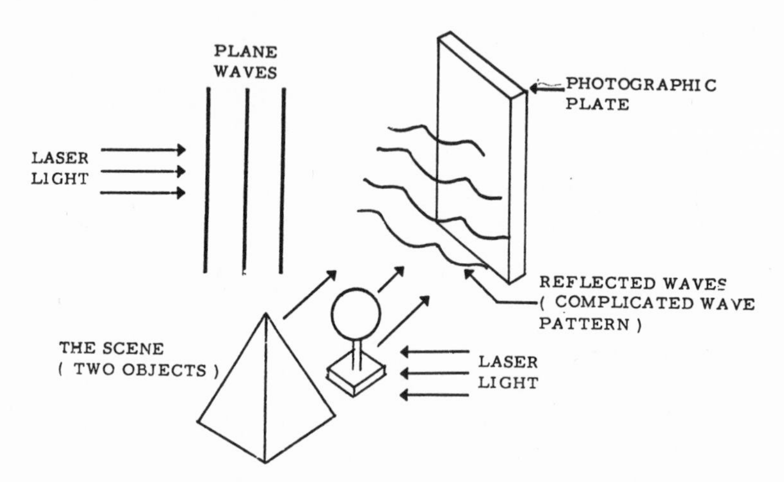

Fig. VI-1. In making a hologram, the scene is illuminated with laser light, and the reflected light is recorded along with a reference wave from the same laser, on a photographic plate. The plate is then developed and fixed.

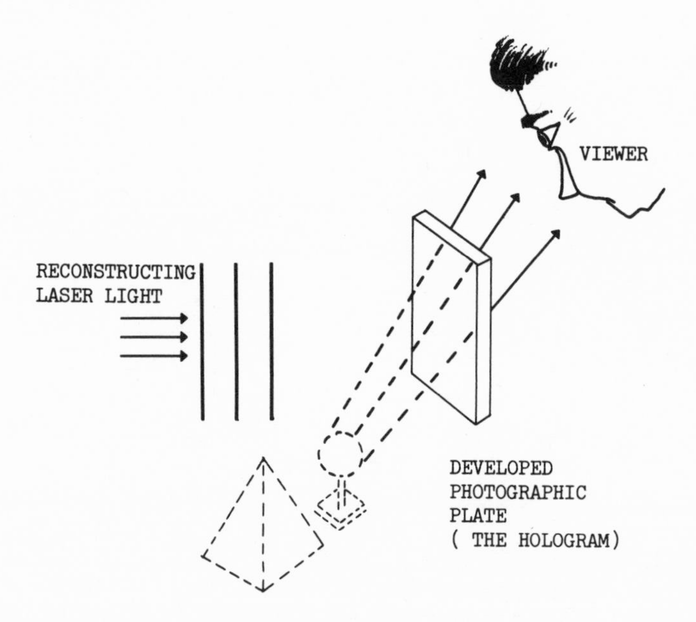

Fig. VI-2. When the developed plate of Fig. VI-1 is illuminated with the same laser reference beam, a viewer sees the original scene "reconstructed," standing out in space with extreme realism behind the hologram "window."

The two original sets of hologram waves are caused to interfere at the photographic plate, as shown in Fig. VI-1. Here the "scene" comprises a pyramid and a sphere. The objects are illuminated by the same source of single-wavelength laser light that is forming the plane waves at the top of the figure. Because the wave fronts of the set of waves issuing from the scene are quite irregular, the interference pattern in this case is quite complicated. After exposure, the photographic plate is developed and fixed, and it thereby becomes the hologram. When it is illuminated with the same laser light used earlier as the reference wave, as shown in Fig. VI-2, a viewer imagines he sees the original two objects of Fig. VI-1 in three dimensions.

To understand how such a light-wave interference pattern, once photographically recorded and then developed, can later recreate a lifelike image of the original scene, let us first examine the very sim-

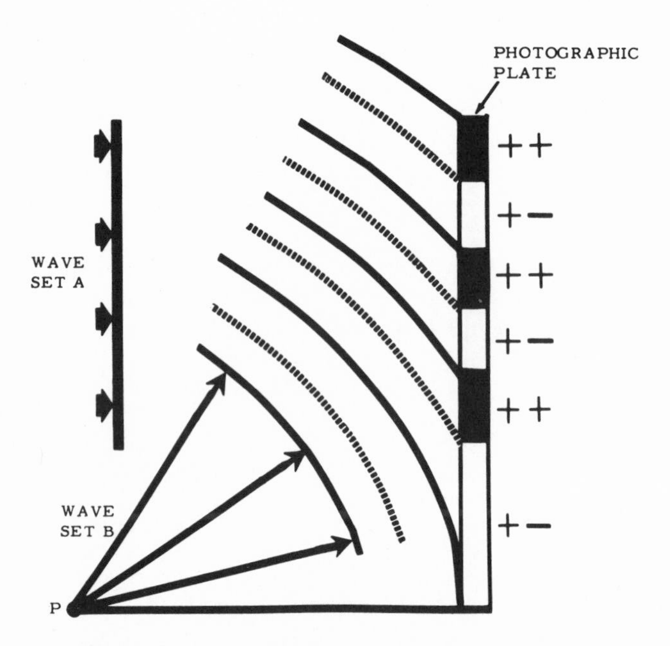

Fig. VI-3. When two sets of single-wavelength waves, one plane and the other spherical, meet at a plane, a circular interference pattern results, with the separation of the outer circles continually decreasing.

ple interference pattern formed when a set of plane waves and a set of spherical waves interfere. This pattern is a circular one having a cross section that is sketched in Fig. VI-3. Parallel plane waves (set *A*) are shown arriving from the left, and they interfere at the photographic plate with the spherical waves (set *B*) issuing from the point source *P*. This interference causes areas of wave subtraction and addition to exist, and, as the distances increase from the central axis, the separation between these areas lessens. The combination of plane and spherical waves generates interference rings.

A Photographic Zone Plate

A photographic recording of the interference pattern between plane and spherical single-frequency light waves is shown in Fig.

VI-4. Now it turns out that the spacings of the rings of both Figs. VI-3 and VI-4 are identical to those of a zone plate. The similarity between holograms and zone plates was first noted by the British scientist G. L. Rogers in 1950. Sections located near the central top and central bottom edges of this pattern resemble somewhat (except for the curvature) the horizontal line pattern of an optical grating.

When an optical grating is illuminated by horizontally traveling, single-frequency plane waves, much of the light-wave energy passes

Fig. VI-4. At the plane of the photographic plate of Fig. VI-3, the wave interference pattern is a series of bright and dark circles. This figure portrays a photographic recording of such a pattern.

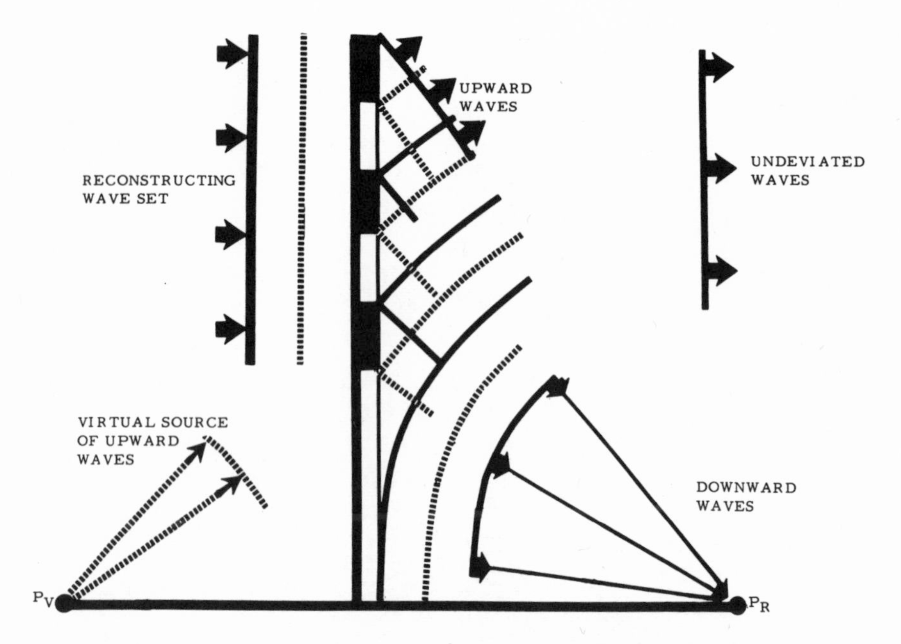

Fig. VI-5. When the upper portion of the circular line pattern of Fig. VI-4 (as recorded in the process sketched in Fig. IV-3) is illuminated with the original, horizontally traveling set of plane waves, three sets of waves result. One set travels straight through horizontally; another acts as though it were diverging from the source point of the original spherical waves; and the third is a set that converges toward a point on the opposite side of the recorded circular pattern.

straight through the grating. In addition, some of the wave energy will be deflected (diffracted) both upward and downward. Accordingly when the patterns of Figs. VI-3 or VI-4 are illuminated with plane waves, three wave sets are similarly generated.

This effect is shown in Fig. VI-5, which portrays the reconstruction of the hologram formed in Fig. VI-3. A portion of the reconstructing plane-wave set arriving from the left is undeviated, passing straight through the photographic transparency. Because the circular striations near the top of the drawing act like the horizontal lines of a grating, energy is diffracted both upward and downward. However, the pattern of striations is circular, so that the waves which are diffracted in the upward direction travel outward as circular waves fronts, seemingly originating at the point P_v. These waves form what is called a *virtual image* of the original point light source P of Fig. VI-3 (virtual, because in this reconstruction no source really exists there). These waves provide an observer, located where the words "upward waves" appear in Fig. VI-5, with the illusion that an actual point source of light exists there, fixed in space behind the photographic plate no matter how he moves his head. Furthermore, this imagined source exists at exactly the spot occupied by the original spherical-wave light source used in making the photographic record, the hologram.

A third set of waves is also formed by the spherical recorded interference pattern. In Fig. VI-5 this set is shown moving downward, and these waves are focused waves. They converge at a point that is located at the same distance from the far side of the photographic record as the virtual source is from the near side. The circular striations of the photographic pattern cause a *real image* of the original light source P to form at P_r (*real*, because a card placed there would show the presence of a true concentration of light).

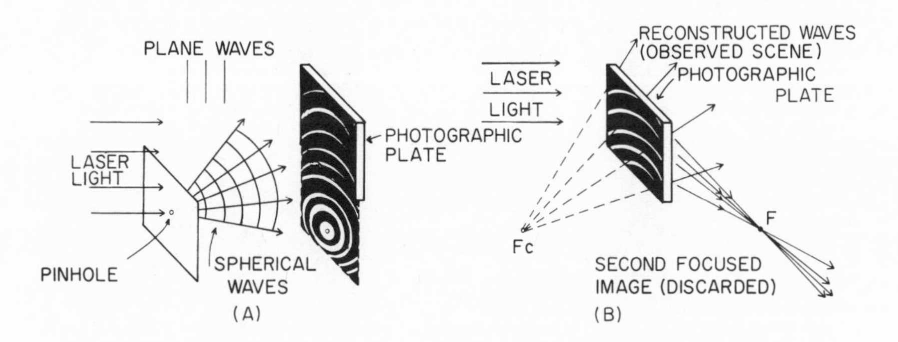

Fig. VI-6. Plane reference waves interfering with spherical waves issuing from a pinhole from a zoneplate interference pattern (A), which, when photographically recorded and reilluminated with laser light, generate waves appearing to emanate from the original pinhole (B).

The Complete Hologram Process

The complete, two-step hologram process is shown in Fig. VI-6. Here, a pinhole in the opaque card at the left serves as the "scene"; it is a point source of spherical waves. These interfere at the photographic plate with the plane waves arriving from the left. In this case only the upper portion of the circular interference pattern is photographically recorded. When the photographic plate is developed and fixed and then placed in the path of plane light waves as shown in the diagram on the right, a virtual image of the original pinhole light source is formed at the *conjugate* focal point, F_c. A viewer at the upper right thus imagines he sees the original light from the pinhole. The real image (the focused image) appears at the true focal point as shown; in the usual viewing of a hologram,

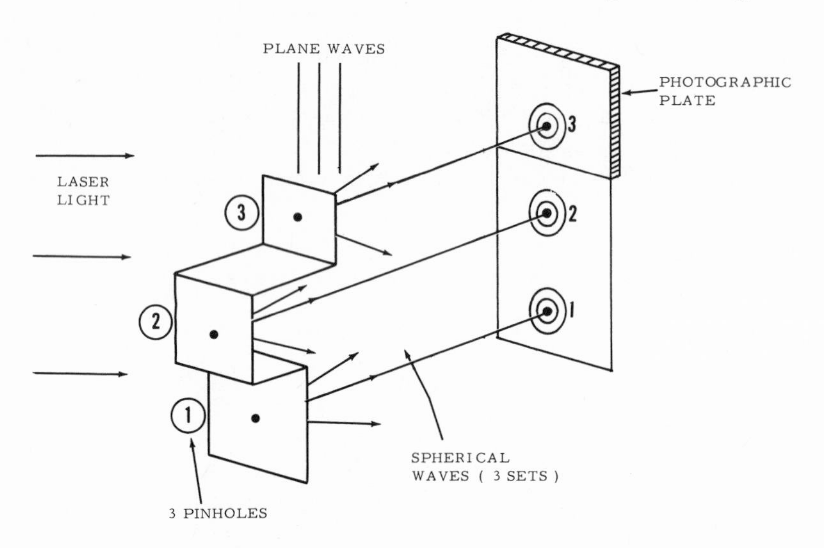

Fig. VI-7. If the single pinhole of Fig. VI-6 is replaced by three separated pinholes, the zone plate patterns of all three are photographically recorded; when this photograph is reilluminated, all three pinholes are seen in their correct three-dimensional positions. A more complicated three-dimensional scene can be considered as many point sources of light, each generating, on the hologram plate, its own zone plate; each of these zone plates will then reconstruct its source in its original three-dimensional position.

this second wave set is not used. In this figure, the straight-through, undeviated waves are not shown.

In Fig. VI-7 a similar photographic recording procedure is sketched, except that for this case, the original scene is one having not a single pinhole, but three pinhole sources of light, each in a different vertical location and each at a different axial distance from the plane of the photographic plate. We see that each of the three light sources generates its own circular, many-ring pattern, comparable to the single pattern of Fig. VI-6. (In the figure, only the first two central circular sections of these three patterns are indicated.) The upper portions of the three sets of circular striations

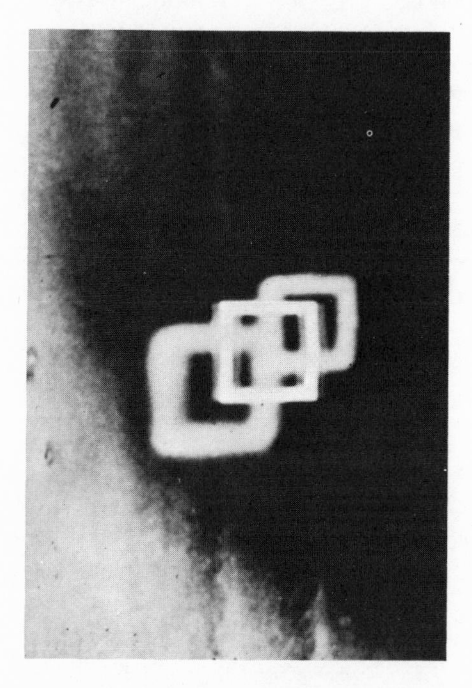

Fig. VI-8. A two-inch diameter hologram zoneplate forming three images. The zone plate is the dark area with a circular outline, and the object was a white square. The zero-order (straight-through) image is the bright square; the virtual, diverging image is to its lower left, and the real, converging image is to its upper right.

(those encompassed by the photographic plate) are recorded as three superimposed sections of zone plates. When this film is developed and fixed and then reilluminted, as was done for the single pinhole recording of Fig. VI-6, three sets of upward waves and three sets of focused waves are generated. Of particular importance from the standpoint of holography is that a virtual image of each of the three pinholes is generated (by the upward, diverging waves). These virtual images cause a viewer at the top right to imagine that he sees three *actual* point sources of light, all fixed in position and each positioned at a different (three-dimensional) location in space. From a particular viewing angle, source number three might hide source number two. However, if the viewer moves his head sideways or up or down, he can see around source number three and verify that source number two does exist.

Most applications of zone plates exploit only their focusing ability. The fact that a zone plate also causes a set of *diverging* waves to be generated is not very well known. But in a hologram this negative lens property of a zone plate is very important for, as has been noted, it is the diverging waves that give the viewer the striking, three-dimensional view of the original scene. Figure VI-8 shows the triple image effect produced by the straight-through, converging, and diverging waves generated by a zone plate.

The Hologram of a Scene

All points of any scene that we perceive are emitting or reflecting light to a certain degree. Similarly, all points of a scene illuminated with laser light are reflecting light. Each point will have a different degree of brightness; yet, each reflecting point *is* a point source of laser light. If a laser reference wave is also present, each such source can form, on a photographic plate, its own circular interference pattern in conjunction with the reference wave. The superposition of all these circular patterns will form a very complicated interference pattern, but it will be recorded as a hologram on the photographic plate, as was sketched in Fig. VI-1. When this complicated photographic pattern is developed, fixed, and reilluminated, reconstruction will occur and light will be diffracted by the hologram, causing all the original light sources to appear in their original, relative locations, thereby providing a fully realistic three-dimensional illusion of the original scene.

The hologram plate itself resembles a window, with the imaged scene appearing behind it in full depth. The viewer has available to him many views of the scene, and to see around an object in the foreground, he simply raises his head or moves it to the left or right. This is in contrast to the older, two-photograph stereo pictures that provide an excellent three-dimensional view of the scene, but only one view. Figure VI-9 shows three photos of one (laser-illuminated) hologram; they are three of the many views a viewer would see if he moved his head from right to left while observing the hologram. For these three photos, the camera taking them was similarly moved from right to left, fully exposing, in the left-hand photo, the original three bars positioned one behind the other.

Fig. VI-9. Three photographs of a hologram being illuminated with laser light. The hologram recorded a scene comprising three vertical bars; for these photos the camera was moved successively farther to the right, finally causing the rear bars to be hidden by the front bars.

Parallax in Holograms

In viewing a hologram, the observer is usually encouraged to move his head sideways or up and down so that he may grasp its

Fig. VI-10. A photograph of a hologram, showing the top of a cut-glass toothpick holder, a carved silver object, and a "thumb-print" glass. The viewer, as he moves his head, will see changes in the light reflected from these objects' surfaces.

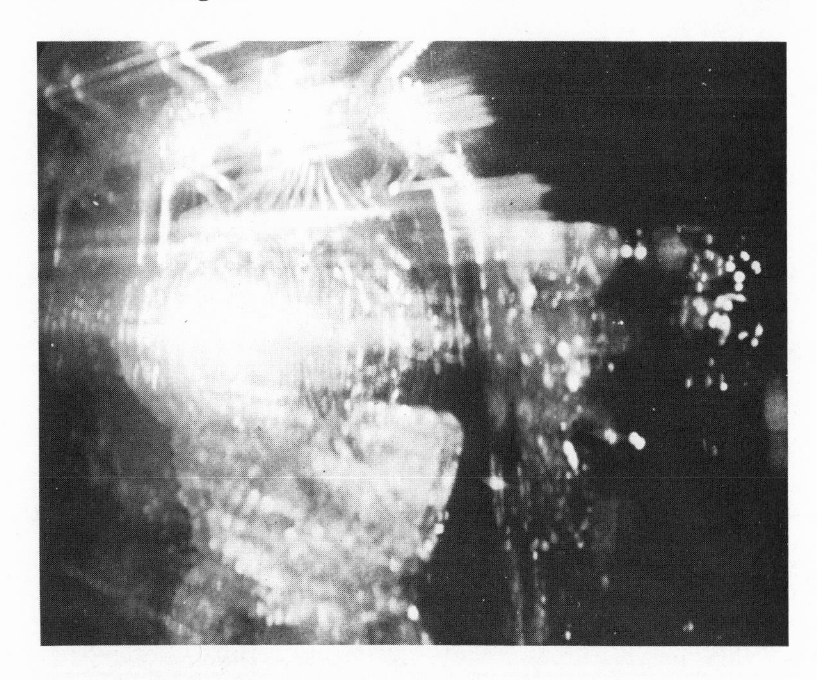

Fig. VI-11. Illuminating the hologram of Fig. VI-10 with a mercury arc lamp results in objectionable blurring.

full realism by observing an effect called *parallax*. In real scenes, more distant objects appear to move with the viewer, whereas closer objects do not. Such effects are very noticeable to a person riding in a train; the nearby telephone poles move past rapidly, but the distant mountains appear to move forward with the traveler. Similarly, the parallax property of holograms constitutes one of their most realistic aspects.

Because hologram viewers invariably do move their heads to experience this parallax effect, hologram designers often include cut-glass objects in the scene to be photographed. In the real situation, glints of light are reflected from the cut glass, and these glints appear and disappear as the viewer moves his head. This effect also occurs for the hologram, and further heightens the realism, as is the hologram of Fig. VI-10.

Single-Wavelength Nature of Holograms

One of the basic properties of ordinary holograms (and zone plates) is their single-frequency nature. Because the design of a zone plate is postulated on one particular wavelength, only waves of that wavelength will be properly focused. Inasmuch as holograms are a form of zone plate, they, too, suffer from this problem; only single-frequency light waves can properly reconstruct their recorded images. If light comprising many colors is used in the reconstruction process, the various colors are diffracted in different directions, and the picture becomes badly blurred (Fig. VI-11).

Nonoptical Holograms

Because the basic feature of holography is that of recording a wave interference pattern, holograms can be made using waves of

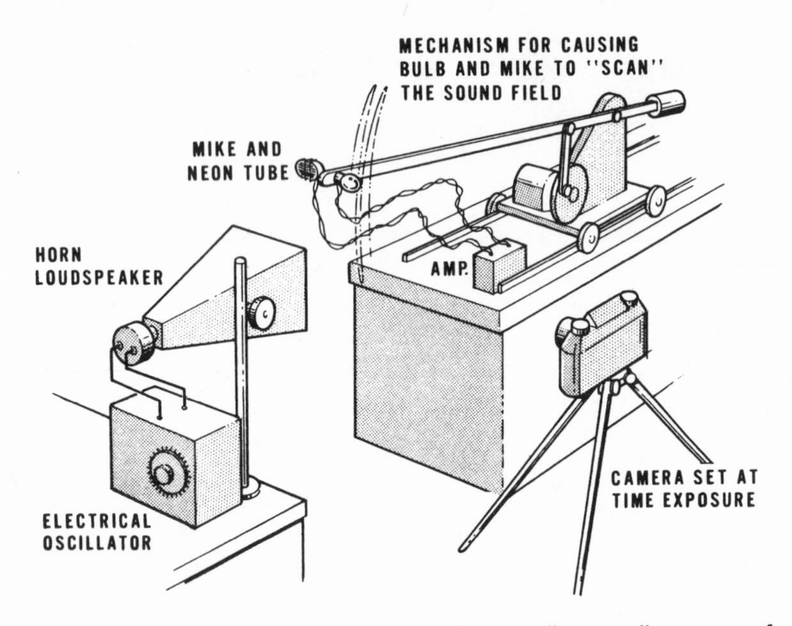

Fig. VI-12. A single microphone-light combination "scanning" an area of interest can record on film the space pattern of sound intensity generated by a horn loudspeaker.

almost any kind. Thus, two sets of radio waves, or two sets of sound waves, can be made to interfere, and if records are made of such patterns, they will be radio (or microwave) holograms, or acoustic holograms.

What were probably the first such holograms to be made were generated as a means for presenting wave patterns, such as those of Figs. I-5, II-3, and II-4. Let us start with a discussion of how Fig. II-3 was made. This is a sound-wave pattern and was recorded as shown in Fig. VI-12. A microphone was caused to scan the sound field, with its amplified output controlling the brightness of a small neon tube moving with the microphone. A camera set for a time exposure could thus provide a visual record of the sound intensity.

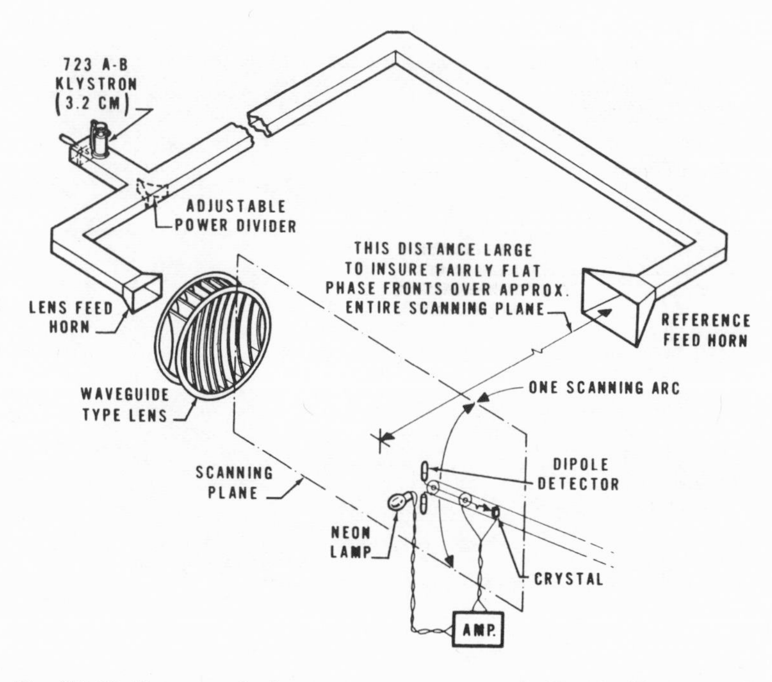

Fig. VI-13. Two sets of coherent microwaves generated by the klystron at the upper left (one set being an off-angle reference wave) are caused to interfere at the scanning plane. The microwave interference pattern is converted into a light pattern by the scanning mechanism, and it is recorded by a camera set at time exposure.

Microwave Holograms

We have seen that to form a hologram, reference waves are needed. Such a reference for the microwave case can be provided as shown in Fig. VI-13. Here the microwaves passing through the lens at the left are the waves of interest, and a set of coherent waves from the reference feed horn at the right interfere at the scanning plane. Again a camera can record this interference pattern and the result is shown in Fig. VI-14. The lines seen are interference lines or microwave "fringes," and this record can thus be classed as a microwave hologram. It was made at the Bell Telephone Laboratories in 1951 by the author and his colleague Floyd K. Harvey. Figure I-5 is a similar record.

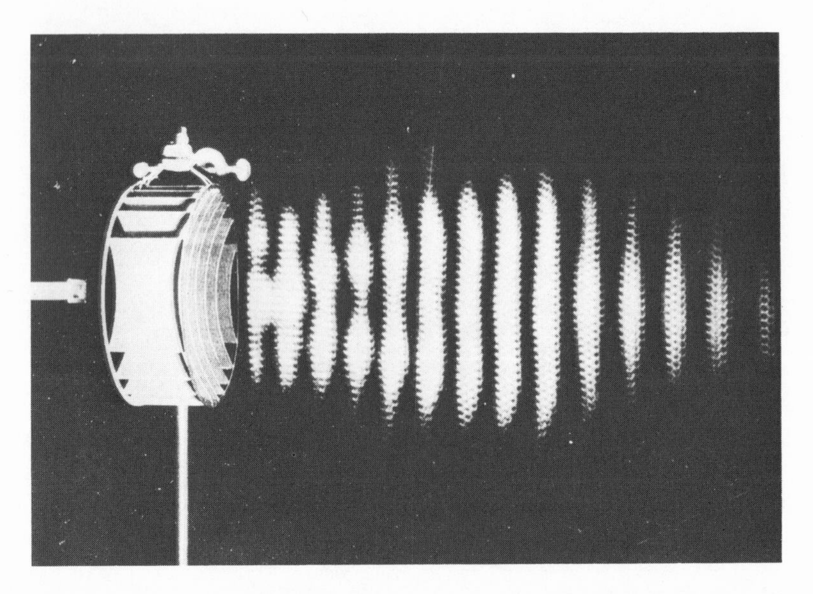

Fig. VI-14. The spherical waves issuing at the left from the same waveguide as in Fig. I-6 are converted into plane waves by a microwave lens. This photo, made as indicated in Fig. VI-13, is a recording of the interference pattern formed by a wave set of interest and a reference wave; it can accordingly be considered to be a microwave hologram, the white vertical striations being microwave *fringes.*

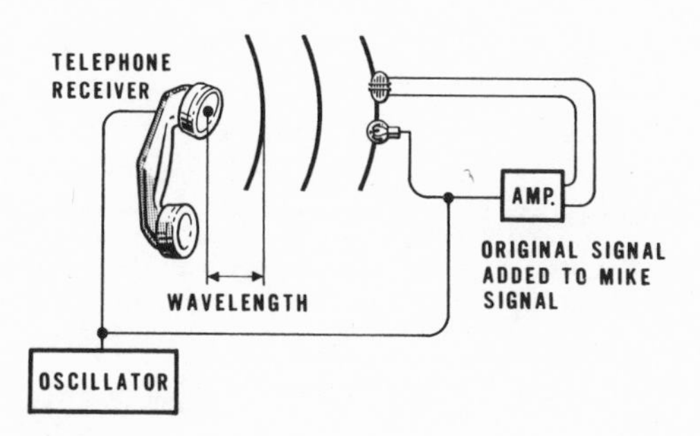

Fig. VI-15. To make an acoustic hologram, the reference wave can be "injected electronically" by means of a connection directly from the oscillator. Figure II-2 was made in the way shown in this figure.

Acoustic Holograms

The telephone receiver pattern of Fig. II-3 is also an interference or "fringe" pattern (that is, a hologram), in this case, an acoustic one. Because sound waves can be transformed so easily into varying electrical currents (and vice versa), a different way of injecting the hologram reference wave was used in making the record of Fig. II-3. The process is shown in Fig. VI-15. A signal, coming directly from the electrical oscillator (which is causing the receiver to vibrate and thus radiate sound waves) is combined with the signal picked up by the microphone. At the junction point of these two sources interferences will occur, and the brightness of the light will again be affected by the constructive and destructive interference effects. The "electronically injected" reference wave is identical in action to that of a set of plane reference waves, and an acoustic hologram results (Fig. II-3).

Microwave Holograms and Liquid Crystals

Microwave interference patterns can also be portrayed using the recently developed "liquid-crystal" technique, in which crystals whose colors are determined by the varying temperature effects introduced by the variations in the strength of the microwave field are employed. Figure VI-16 shows such a liquid-crystal microwave interference pattern generated by two interfering coherent microwave sources. One was a point source and one approximated a plane-wave source, and the two wave sets were made to interfere at the plane of the liquid-crystal device. The resulting pattern is identical to the pattern of an offset microwave zone plate. Since this zone plate is a hologram, the reconstruction of an image (either real or

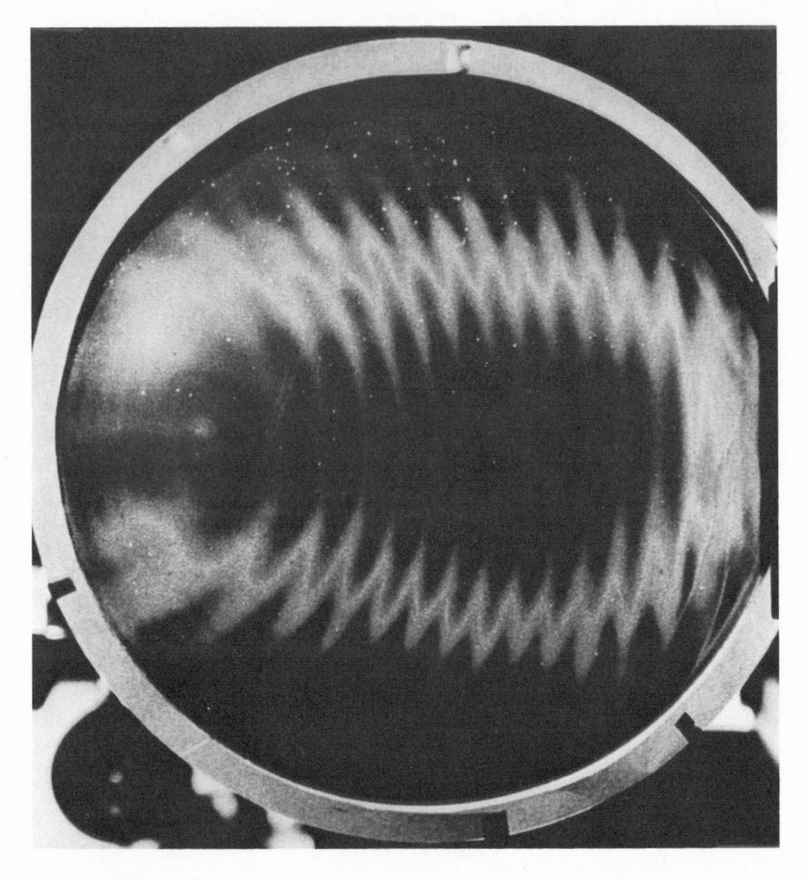

Fig. VI-16. A liquid-crystal pattern formed between a point microwave source and a plane-wave source. Because it is offset, the straight-through (zero-order) component and the two images can be separated.

virtual) of the original point microwave source can be accomplished optically by photographically reducing the record and illuminating it with coherent (laser) light.

Had there been a large number of microwave point sources in the original microwave "scene," each would have formed, in conjunction with the reference wave, its own two-dimensional zone plate, and the liquid crystal pattern would have been, as in an optical hologram, a superposition of a large number of zone plates. By photographically recording, reducing, and reilluminating this pattern with laser light, a three-dimensional, visual portrayal of the many microwave sources would have been formed.

Ultrasonic Holograms

We have noted that two-dimensional acoustic holograms like the liquid crystal hologram of Fig. VI-16 can be formed by recording a sound-wave interference pattern (Fig. II-3). When actual objects comprise the "scene" to be "illuminated" with sound waves, the wave interference pattern is first transformed into a light-wave pattern that, after a size reduction, is viewed with laser light for optical reconstruction of the scene.

Dr. Rolf Mueller of the Bendix Research Laboratories has pioneered in using, as an ultrasonic hologram surface, a liquid–air interface (Fig. VI-17). A coherent reference wave is directed at this liquid surface, and it becomes the "recording" surface for the hologram interference pattern. Because the surface of a liquid is a pressure-release surface, that is, a surface that "gives," or rises, at points where higher-than-average sound pressures exist, the acoustic interference pattern transforms the otherwise plane liquid surface into a surface having extremely minute, stationary ripples on it. When this rippled surface is illuminated with coherent (laser) light, an image of the submerged object is reconstructed.

Underwater Viewing

Acoustic holography when fully developed should be a valuable tool for such underwater viewing activities as naval search and

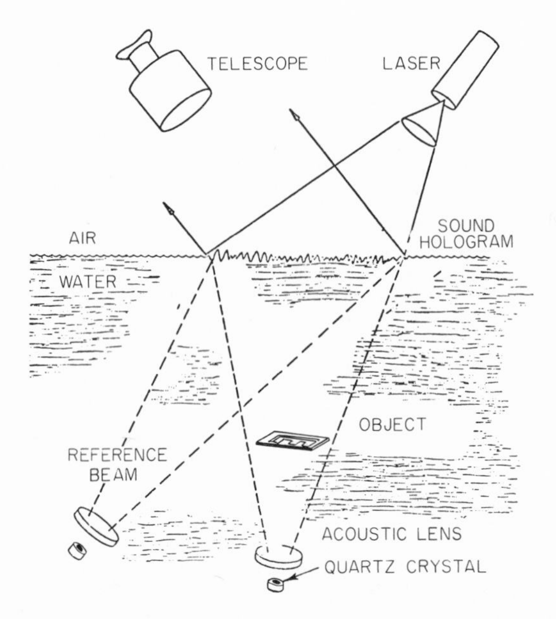

Fig. VI-17. When the waves of interest and the reference waves of an underwater acoustic hologram interfere at the water surface, the ripples constitute the hologram, and immediate reconstruction can be brought about by shining laser light on these interference ripples.

surveillance, detection and identification of schools of fish, and control of harbor traffic. Techniques currently available for these purposes are limited to sonar and optical viewing. Optical techniques are limited, sometimes severely, by the turbidity of the water.

Figure VI-18 presents an artist's concept of an underwater viewing system for use in a search vehicle of the future. The transmitters and receivers may be combined into a single array. The system may also be used as a standard sonar range-detection system.

Earth Exploration

The application of holographic techniques to large-scale underground viewing is still in initial development stages but promises to aid in the accumulation of geological information for scientific pur-

Fig. VI-18. Underwater viewing using acoustic holography.

poses and to have important implications for the mining and oil industries. Object *sizes* in work of this type are several orders of magnitude greater than those in the ultrasonic applications just discussed and object *distances* are measured in hundreds of meters. For good penetration of the geological structures, the frequency range to be employed must range between 10 and 100 Hz. Frequencies so low (and wavelengths so great) place unusual constraints on detector arrays, but scanning techniques could be applicable in this work, since the objects under study are immobile.

The basic elements of a system for locating offshore oil deposits are depicted in Fig. IV-9. A cable which in practice would be 100 wavelengths or more in length, is towed behind a ship equipped with a high-power transmitter capable of emitting low-frequency coherent acoustic energy into the ocean depths. Signals reflected or scattered from the ocean bottom or from the geological layers below are picked up by the cable array. Holographic processing of seismic data obtained from large arrays of scanned hydrophones should permit the retrieval of useful information from such acoustic signals.

The Concept of Phase in Holography

We have seen in the previous chapter, in discussing phased-array radars, that the concept of phase is quite useful in sonar and radar. When an array is "phased" in one particular way, its beam points in one direction, and when phased in another way, the beam points in another direction. One procedure often followed in the case of passive receiving arrays is that of incorporating large numbers of elemental delays or phase shifts. The individual elemental receivers are connected in a multitude of different ways, so that many receiving beams, all pointing in different directions, are effective at all times. The array then "looks" in many directions at once.

Because a photographic plate of film is sensitive only to the intensity of the light falling on it, it is often said that ordinary photographs record only the intensity or amplitude, not the phase of the light field, and we shall see that strictly this is quite correct. Because the radar and sonar engineer utilizes phase and phase shifts as he "phase steers" his arrays, he normally looks upon a receiving array that can thereby "look" in many directions at once, as taking *phase* information fully into account. The phasing of his array elements must be done completely and precisely in order to provide all beams with maximum effectiveness.

Now a passive sonar or radar "looking" in many directions is practically identical to a camera that looks in many directions at once, recording on its film the light strength arriving from all these directions (just as the passive sonar measures the sound intensity arriving from many directions at once). But as we noted above, the camera film can only record the intensity of the light pattern generated at the plane of its emulsion; it does not record the *relative* phases of various separated areas of the pattern. Accordingly, a further refinement of the definition of phase is needed.

Let us think once again about the light pattern at the plane of the emulsion of the camera film. It unquestionably does have variations in intensity, which provide, through the usual photographic process, a picture of the scene. But there is also phase information in that light pattern. For example, the waves responsible for two different but equally bright points in the final picture might have been different in (relative) phase by a full 180°, yet both would appear as indistinguishable bright spots in the final picture. A similar out-of-phase situation could exist in the signals received on two sonar beams without affecting in the least the final sonar record.

But the fact that phase variations can and do exist in the light pattern does not yet tell us that this information is important. It took holography to show how useful a record would be that did record both the amplitude and the (relative) phase of the wave pattern. First a way had to be found to record these two features of the pattern, and Gabor achieved this by his use of a reference wave. The plane reference waves can clearly indicate the situation in the example above, since the one bright area having its phase 180° opposed to the other would interfere differently with the reference waves. Obviously, the three-dimensional reconstructions of holograms, not possible with one ordinary photo, proved without a doubt that the phase information *is* useful and that photography can not lay claim as holography can to providing the "complete message."

This important difference between holography and photography (and therefore between ordinary radar and coherent radar) will become more evident in the next chapter.

Chapter VII

COHERENT RADAR AND SONAR

We discussed in the last chapter the fundamental property of holography which shows that, in spite of the similarities, holography is not just a variant of photography, but rather a basically new process. We remarked also that a new form of radar, coherent, or synthetic-aperture, radar, has exploited many of the advantages of holography. Let us therefore examine some of the properties of this new coherent form of radar.

Synthetic-Aperture Radar

The similarity of coherent radar, which requires single-frequency emission for operations, and holography, is apparent in examining the synthetic-aperture aircraft radar procedure. As an aircraft moves along a very straight path, its radar continuously emits successive microwave pulses as shown in Fig. VII-1. The frequency of the microwave signal must be constant for the signal to remain coherent with successive signals for long periods.

During these periods in which the signals remain coherent the aircraft travels several thousand feet, and because the signals are

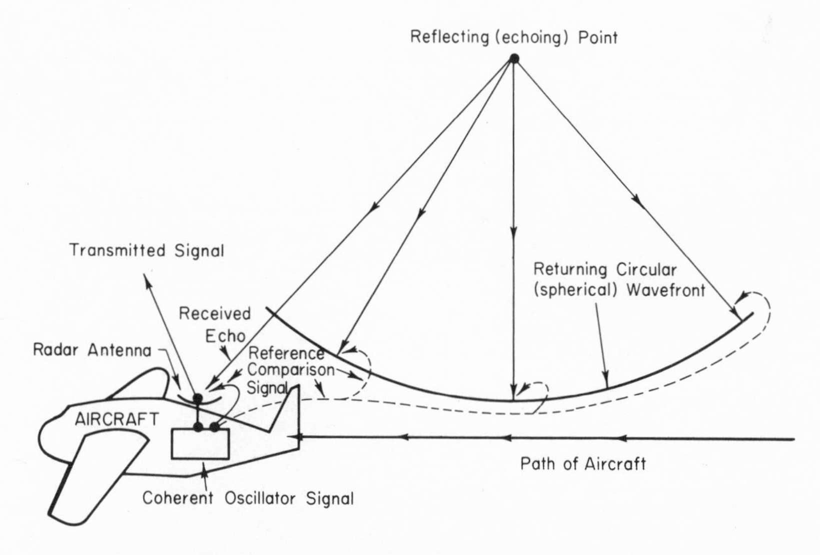

Fig. VII-1. Principle of side-looking radar. Echoes from the reflecting point are received over a considerable interval and assembled by a holographic method that creates interference patterns between the return signal and a reference sampled from the transmitted signal. The result is a synthetic-aperture, high-resolution atenna whose length is that of the flight path.

coherent, the many returned echoes appear to be received by a single antenna. Thus, this antenna has a long synthetic aperture, equal to the distance traveled. This effectively large antenna length provides both the very high gain and resolving power of coherent radar, yielding extremely fine detail as in the aerial photograph, Fig. VII-2.

As in holographic imaging, the microwave generator that provides the illuminating signal also acts as a reference wave. The reflected signals received along the flight path form a complex interference pattern with the reference signal which, photographically processed and superimposed, becomes a record of the area of interest. Single echoing objects generate one-dimensional zone plates, just as in the first stage of holographic reconstruction. Thus the photographic record of the echoes received by coherent radar is a microwave hologram.

How the interference pattern is formed is illustrated in Fig. VII-1. For simplicity, only one reflecting point is sketched. Reflections returning from this point exhibit spherical wave fronts, while the reference signal acts as a set of plane waves perpendicular to

Fig. VII-2. A processed radar hologram yields high-resolution images. Parts of Detroit and Lake Erie are shown in this radar reconstruction.

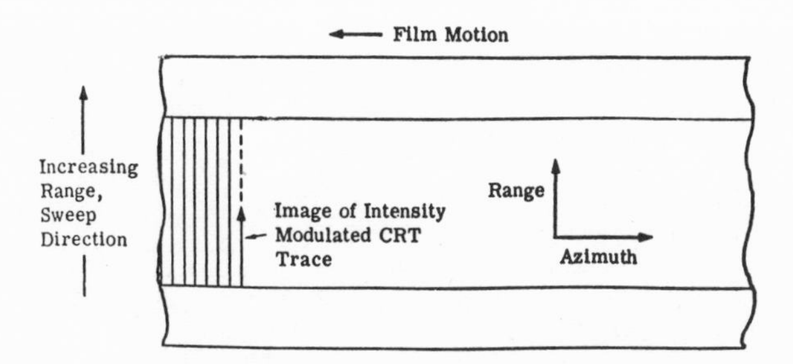

Fig. VII-3. A hologram record in side-looking radar is formed by photographing an intensity-modulated cathode-ray tube trace.

the airplane's path. The combination of the received signal and the coherent reference signal is amplified and modulates the intensity of a cathode-ray tube trace (Fig. VII-3). Each vertical line represents signals received from all range points; the points at greater range are recorded nearer the top of the vertical trace. As the aircraft moves and new pulses are emitted, the film is indexed (moved slightly to the left) to record a new set of returns—a new vertical line.

When there is only one reflecting point at a given range, the upward-moving, cathode-ray beam would be brightened for every pulse only at that one point in range—at the same level on each vertical trace. The echoes thus would be recorded as a single hori-

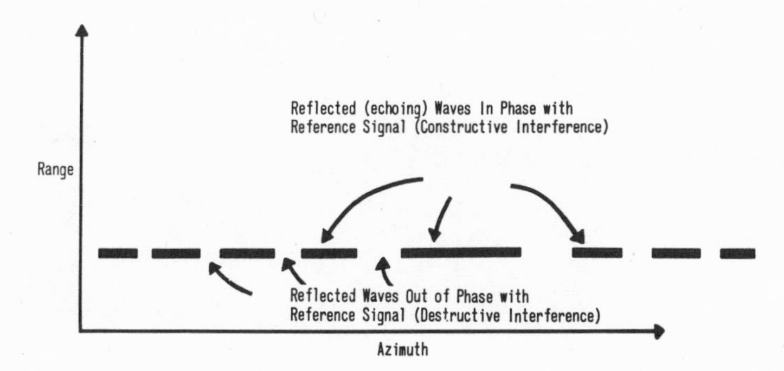

Fig. VII-4. A photographically stored hologram of a single point: the recorded signals form a zone plate.

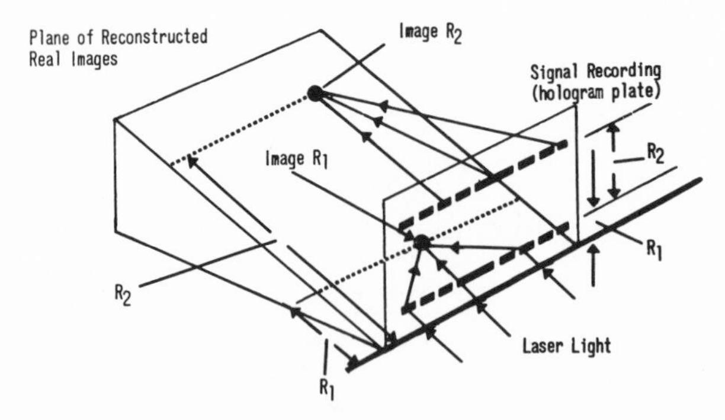

Fig. VII-5. Reconstruction by laser light of a side-looking radar hologram.

zontal line. But this line is not continuous, as shown in Fig. VII-4. Because the returning waves are circular, different points on the wave reach the receiver at different times, so that the combination of the returning signal and the reference waves successively produces constructive and destructive interference areas. And because of the circular nature of the waves, the slant azimuth range (the distance from the aircraft to the reflecting point) changes.

At the greater slant angles, where the distance from the plane to the point changes rapidly, this succession of in-phase and out-of-phase conditions occurs rapidly. But when the aircraft is nearly abreast of the reflecting point, the range is almost constant, so the changes come more slowly. The resulting record is a one-dimensional zone-plate hologram, which, if illuminated by laser light, re-

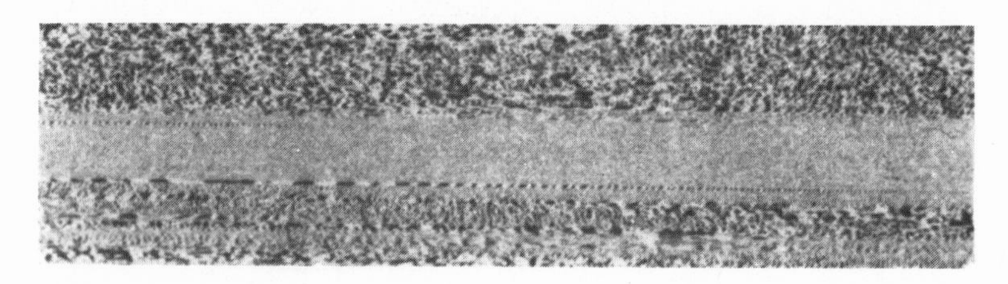

Fig. VII-6. A hologram record produced by side-looking radar. Note the zone-plate signature of a particularly distinct reflecting object at the lower portion of the blank area.

constructs the reflecting point just as illuminating a hologram reconstructs an image of the original scene.

Plane-wave laser light (Fig. VII-5) reconstructs the range and azimuthal position of the reflecting points that generated the zone plates. The two reflecting points shown are displaced appreciably in range and slightly in azimuth. The reconstructed images fall on a plane; the tilt of the plane is determined by the radar's vertical tilt. Figure VII-6 is a portion of an actual radar record made in a synthetic aperture radar. An isolated reflecting point caused the one-dimensional zone plate seen at the bottom of the central blank area.

Many people first questioned whether high resolution could be maintained for nearby objects. Because of the synthetic-aperture an-

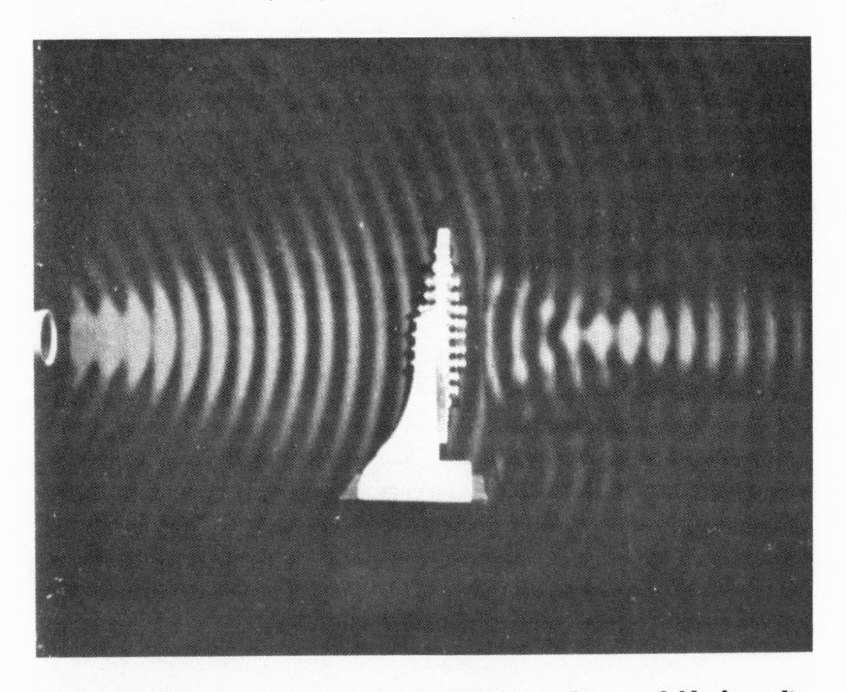

Fig. VII-7. To cause a concentration of energy in the near field of a radiator, the outgoing wave fronts must be curved and centered on the near field point. Here an acoustic lens provides the needed phase shifts (time delays) for the waves arriving at the left so as to concentrate the sound energy in the near field of the lens aperture.

tenna's great length, the majority of reflecting objects are not in the far field, where most conventional radar antennas operate, but in the near field, as was shown by Dr. Louis Cutrona and his group at the University of Michigan.

For near-field reflection, radars with large-aperture antennas can achieve maximum efficiency and resolution only when the phases of the antenna elements are adjusted so that the beam pattern forms an arc of a circle centered on the object. Yet for other near-field targets located at other ranges and azimuth directions, the phases of all antenna elements similarly must be adjusted to correspond to the arcs of circles centered on these other points. Such a result would be difficult to achieve with standard phased-array radar procedures.

This effect can be explained by reference to Fig. VII-7. Here an acoustic lens has introduced phase shifts (time delays) into the wave pattern arriving at the left of the lens, thereby causing the waves proceeding on toward the right of the lens to have circular wave fronts converging on a focal area (the bright region at the right of the photo). This area is in the "near field" of the lens.

We saw conversely, in connection with Fig. I-12 (involving flat wave fronts, such as those shown in Fig. I-6), that no concentrating effect existed in the near field. The antenna resolution and gain for near-field objects for a radar designed along those lines would be very poor indeed.

In the reconstruction of the radar records, as shown in Fig. VII-5, (one-dimensional) zone plates provide the lens effect of Fig. VII-7, and cause the laser light to be similarly concentrated. Reconstruction of all of the near-field reflecting objects thereby provides a highly detailed photo (Fig. VII-2) of the original terrain.

Thus, once it is recognized that a coherent radar record is a hologram, it becomes apparent that high resolution in the near field can be accomplished without complex phased-array techniques. In holography each small light-reflecting point generates its own zone plate, and each of these is then used to reconstruct each point in space. Similarly, synthetic-aperture radar photographically captures the curved wave fronts emanating from each reflecting point by combining them with a reference wave, generating many one-dimensional zone plates.

In reconstruction, diffraction of a coherent laser beam by the zone plates produces the properly curved wave fronts in the beam, so that the light will be focused at points corresponding to the reflecting targets in the original landscape. Just as in an optical hologram, each point of a three-dimensional scene is brought into sharp focus at any distance from the photographic plate, so the microwave hologram of the radar record provides good focus on all its reconstructed points.

This capability is common to both coherent radar and holography, and sets them apart from ordinary radar and photography. Usual optical imaging processes employ lenses or paraboloidal reflectors that permit only one plane section of the image field to be recorded in truly sharp focus, a limitation that does not apply to holography or synthetic-aperture radar.

However, there is one inherent processing limitation in holographic radars: A photographic record must first be made, and this record, after the film is developed, must be illuminated with coherent light. The delay involved cannot meet the real-time requirements of many applications. But real-time readout *can* be achieved without film by using ultrasonic cells. These can modulate (diffract) a coherent light beam illuminating the cell in almost the same way a photographic film does. When the entire microwave–acoustic signal has moved into the ultrasonic cell, it is briefly illuminated with laser light, and the resulting diffraction pattern provides the target information.

Stationary Hologram Radars

Still broader uses can be made of hologram radar concepts, particularly when large apertures are involved. In side-looking radar, Doppler effects are introduced into the received signals by the airplane's motion. However, in forming holograms, relative motion, and hence Doppler effects, are absent; thus, through holographic methods, stationary coherent radars become feasible. Such radar antennas could be extremely long linear arrays of independent receivers. Other possibilities include crossed arrays (Mills Crosses) and square arrays.

It should be noted that in holography, even a small portion of the hologram is able to reconstruct the full image. Similarly, in stationary coherent radars, just the end sections of the long linear arrays, or the four-corner sections of square arrays, could be used to maintain maximum resolution.

Probably the simplest form of a stationary coherent radar antenna would be a long linear array of elements identical to the aircraft-borne antenna described earlier. In operation, each element would first transmit a short pulse, following which it would act as a receiving element. The process is repeated, with the transmitted signal successively transferred from one antenna element to the next, matching the aircraft motion to be simulated. Small amounts of coherent signal must be fed continually to each element to provide the holographic reference signal. The returning echoes, interfering at each receiver element with this reference wave, would produce an interference pattern. The returning signals then are recorded on film, indexed each time the transmitted signal is passed on to the next antenna element.

Bistatic Coherent Radar and Sonar

In addition to side-looking radar, hologram concepts also can be extended to forward-scatter, bistatic systems. Here the targets are located between the transmitter and a physically separate receiving array, taking advantage of the high forward-scatter signal diffracted by targets. The receiving array again could be either one long linear array of many elemental receivers or two end sections of that array.

For a (stationary) single-point scatterer, the interference pattern between the scattered signal and the reference signal again would be a zone plate. The receiving array would intercept a linear section of it (a one-dimensional zone plate). By photographically recording the individual outputs of all of the receiving array elements as one photographic line, comparable to the one line of a side-looking coherent radar record, the range and azimuth of this scatterer then could be determined from the one-dimensional zone plate.

With many scatterers, multiple zone plates are generated; these would be photographically superimposed, and positional information on all of the scatters could be retrieved in the usual holographic manner.

Synthetic End-Fire Systems

We saw earlier how end-fire radiators, such as the one shown in Fig. II-23, can generate a sharp beam and hence provide high "gain." Just as side-looking radar achieves the equivalent gain of a very long linear broadside array, the synthetic-aperture procedure can also provide synthetic end-fire gain. In fact the same directivity gain can be achieved as in the linear broadside antenna. Such an end-fire radar technique would endow a small airborne antenna with a very high forward directivity gain. The resulting pencil-shaped beam could enable the aircraft to detect weakly reflecting objects in its path.

In the end-fire case the interference pattern produced by echoes from a front target interfering with a reference wave is a one-dimensional, uniformly spaced, sinusoidally varying density pattern, instead of the nonuniformly spaced, one-dimensional zone-plate pattern of the side-looking radar of Fig. VII-4. This pattern is equivalent to a one-dimensional grating, and it is this grating that is then used to reconstruct the image. The plane waves of coherent light diffracted by this grating are converted by a lens into circular waves converging at a focal point. Light concentrations equal to that obtained with a one-dimensional zone plate are thus possible. A correspondingly high effective antenna gain is also achieved, with its accompanying high signal-to-noise ratio.

The narrowness of the synthetic end-fire beam minimizes reflections from the ground when the aircraft is in flight at high altitudes. Only those targets in the direct path of the plane will return a strong signal. The high synthetic gain thus achieved may permit detection, at useful ranges, of those discontinuities having small indexes of refraction that occur in regions of clear air turbulence. However, real-time reconstruction and display is essential if the aircraft is to take corrective action when the returns indicate turbulence.

Passive Coherent Radar or Sonar

A passive coherent system becomes possible when a target radiates a highly coherent signal. Such radiation could originate from specially equipped aircraft in the vicinity of airports, whose location then could be determined. The airport equipment might comprise one long array of receiver elements, with one or more elements receiving, amplifying, and feeding the aircraft signal *as a reference wave* to the other elements. These elements also would be receiving the radiated signal directly. Thus an interference pattern would be generated along the array, which, following a recording and reconstruction process, would locate the radiating object.

A variation of the procedure is to put a transmitter at the location of interest, say an airport, sending signals to aircraft. These targets would receive the signal, amplify it, and reradiate it. The transmitted signal would be fed, as a reference wave, to the individual receivers of the receiving arrays.

Holographic Pulse Compression

Some attention is being given to the possibility of using holographic methods in pulse-compression radar. Pulse-compression techniques convert long, high-power transmitted pulses to shorter, high-resolution pulses at the receiver. We noted in Chapter II that often the frequency is varied (chirped) over the duration of the pulse and then is detected and compressed by a matched (dispersive) filter. The filter causes signals of different frequencies to travel at different velocities, thereby permitting the more rapidly traveling portion of the long pulse to catch up to the slower, earlier-generated portion.

In a new technique, the *amplitude* of the long pulse is varied instead of its *frequency*, so as to have an envelope corresponding to a zone plate. The total time of the original pulse corresponds to the total length of the zone plate; thus the long CW pulse gets an amplitude-modulation pattern corresponding to the spacing of a one-dimensional zone plate.

As before, the received echoes are photographically recorded, each from a single, isolated target generating its own photographically recorded zone-plate pattern. Optical processing is accomplished by focusing a laser beam onto a tiny focal area, thereby providing a very accurate range for the reflecting point, equal to the zone plate dot spacing that can be resolved. The ratio of this resolution width to zone-plate length corresponds to the "pulse compression ratio." With this method compressions of 5000 to 1 are feasible.

When there are many reflectors at various ranges, many received zone plates would be generated. These would be superimposed on the photographic record, yielding a line recording of numerous superimposed zone plates. This radar record then would be laser processed to yield individual targets.

This amplitude-modulation procedure is particularly adaptable to synthetic-aperture radar where optical processing techniques already are employed. To increase the energy in the outgoing pulse of such radars, the AM pulse can also be made quite long, and be given a shape corresponding to the diffraction pattern (envelope) resulting from the zone plate. This pulse shape will cause the thin,

Fig. VII-8. Top: a sketch of a single zone plate in a radar record (such as shown in Fig. VII-6), caused by one reflecting object. Bottom: a sketch of a record, again generated by one reflecting object, for the situation where a long pulse is substituted for the usual very short radar pulse. The long pulse is given an amplitude envelope corresponding to a zone plate, and optical processing is first used to collapse the vertical dimension of the lower pattern into that of the upper one.

one-dimensional zone plate illustrated in Fig. VII-4 to acquire an extended vertical dimension, thereby becoming a sort of two-dimensional zone plate as shown in the bottom of the Fig. VII-8.

With optical processing procedures, this pattern is collapsed into a one-dimensional pattern for range information. In a second step, the normal holographic processing would be applied to retrieve the azimuth information. With many targets at different ranges and azimuthal positions, zone-plate superposition would occur in both horizontal and vertical directions, but optical processing still would permit full retrieval of the information.

Holographic Processing of Bistatic Radar Signals

The holographic process can be used to enhance CW bistatic radars where weak signal returns are lost in noise. Fringe-area television reception is degraded when an aircraft flies between the transmitting station and the receiver. The signal reflected from the aircraft generates interference effects that cause variations in picture brightness—rapidly at first, then slowly, then rapidly again.

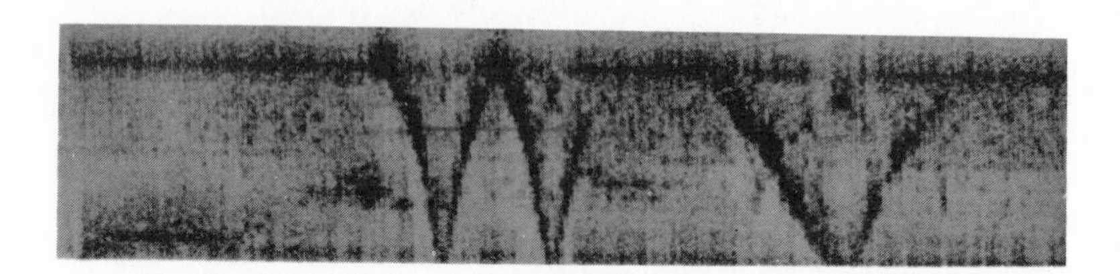

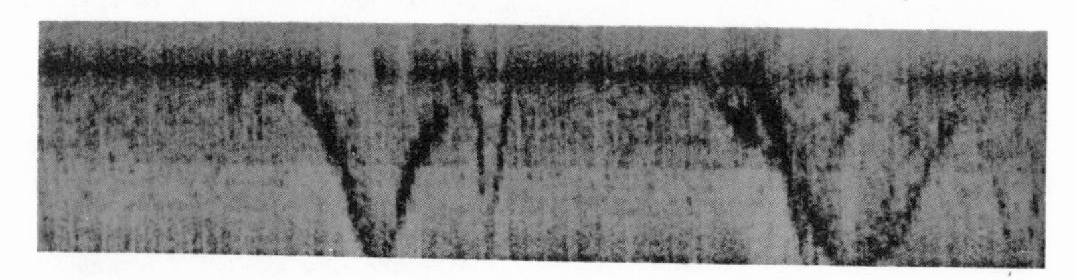

Fig. VII-9. Frequency versus time records of the interference effect caused by aircraft flying over a radio transmission link.

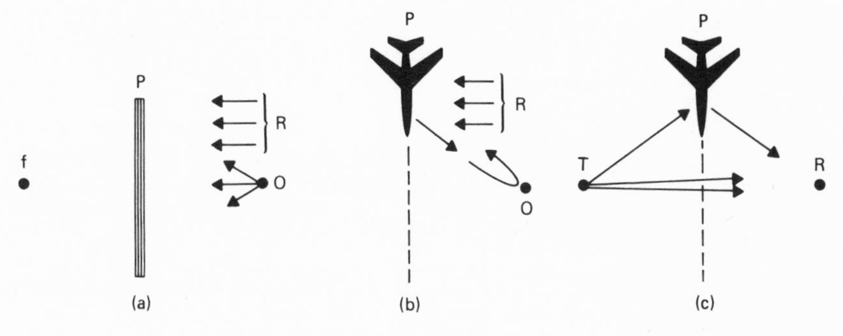

Fig. VII-10. Three similar zone-plate generators. (a) Hologram of a point source. (b) Single reflecting point being recorded as a one-dimensional zone plate by a synthetic-aperture radar. (c) One-dimensional zone plate generated by an aircraft flying over a radio transmission link.

A time–frequency plot of such changing-frequency signals is shown in Fig. VII-9 for a number of passing aircraft. Time is plotted horizontally in both top and bottom records, and frequency vertically. The shapes of the two records at the left of center show that the first plane, which generated the wider V curve, passed by more slowly than the second.

The principle has been applied to a bistatic transmission-line arrangement in a CW radar for observing meteor trails. Constant-velocity targets generate return signals in such radars, which have a time–amplitude pattern closely resembling that of a zone plate. Accordingly, the same coherent optical techniques employed in holography and synthetic aperture radar should improve the receiver's signal-to-noise ratio. This would improve the readability of patterns that are obscure or very weak, like those in the lower right of Fig. VII-9.

The similarity between transmission-path interference signals and certain hologram and synthetic aperture radar records are indicated in the three situations sketched in Fig. VII-10. In (a), a point-source hologram is being recorded on photographic plate *P*. A two-dimensional zone plate results when CW spherical waves from the laser source *O* interfere at the plate with the plane waves of a reference beam *R* generated by the same source.

In (b), a single reflecting object O is recorded on a synthetic-aperture radar record. The plane P, flying along the dotted-line path, periodically transmits short pulses of coherent microwaves, and the signals reflected from the object O combine with a coherent reference signal shown as the set of plane waves R. The interference pattern thus formed again corresponds to a one-dimensional zone plate when photographically recorded.

In (c) the CW signal from the transmitter acts as the reference signal at the receiver, but now reflector P is moving rather than the transmitter and receiver as in (b). The resultant combined signal recorded at R is (for identical aircraft motion) almost equal to the combination signal recorded photographically in the aircraft of situation (b).

Obviously, the individual V curves of Fig. VII-9 also could have been recorded photographically, as in Fig. VII-6, thereby forming one-dimensional zone plates. In this case, each recorded zone plate would have a focal length determined by the plane's speed. Thus, the cluster of undetectable records at the lower right of Fig. VII-9 would be transformed into a set of superimposed zone plates. Standard coherent optical processing procedures then could retrieve the desired information for each target. The chief advantage would be the very large increase in signal-to-noise ratio, equivalent to the very sizable signal-to-noise improvement obtained by synthetic-aperture radar procedures.

Far-Field Hologram Radar

Because both holography and coherent radar are superior to lens optics and ordinary radar in providing detailed information concerning objects located in the near field, one might be tempted to dismiss consideration of the far-field performance of coherent radar as unimportant, since ordinary radar performs this task well. On the other hand, hologram techniques may provide simpler ways of achieving performance equivalent to the usual far-field radars or sonars.

Consider, for example, a multielement, square-array radar, having 100 elements on a side and hence 100×100 or 10,000 total elements. The far field for this array for microwaves of 1-in. wavelength begins at a distance of a little over 400 ft from the array. Such a radar would thus be considered a "far-field" radar. We now suppose that the array itself is used only for receiving, that the transmitted signal is radiated by another antenna, and that the array is to acquire information regarding reflecting objects located within the radiation volume illuminated by the transmitter.

One way of accomplishing the desired function of the array would be by the pulsed-array procedure, for example, by establishing thousands of "pre-formed" beams, whereby, for any single beam direction, delays would be provided as necessary for each of the 10,000 array elements so that all elements would contribute properly to the receiving beam for that direction. For a 90° pyramid of coverage, approximately 8000 pre-formed beams would be required, and for 10,000 elements and 8000 beams, 80,000,000 delay elements would have to be provided.

Consider now providing these same far-field beams by holographic, that is, by stationary coherent radar, procedures. The transmitted signal is then coherent, provided by a highly stable oscillator. A small amount of this oscillator signal would be used as a reference signal and would be supplied continuously to each element of the receiving array so as to simulate a plane reference wave impinging on the entire array. Reflecting objects located within the cone of the transmitter will reflect some of the transmitted signal back to the array, thereby generating, in combination with the reference signal, a holographic interference pattern that is sampled by all the elements. The resultant values at all elements are then recorded photographically, and, after development of the photographic record, the far-field reflecting objects are reconstructed by coherent light.

Even though this process causes only the individual output of each single isolated unit of the 10,000 elements to affect, by itself, the exposure of the photographic plate at that point, the coherence of the signals generating the (zone-plate) interference patterns at the array plane, and the later focusing by these zone plates of the laser reconstructing beam causes the receiving "array gain" thereby achieved to be quite comparable to that obtained in the normal process where all elements are paralleled (with proper phasing for each beam).

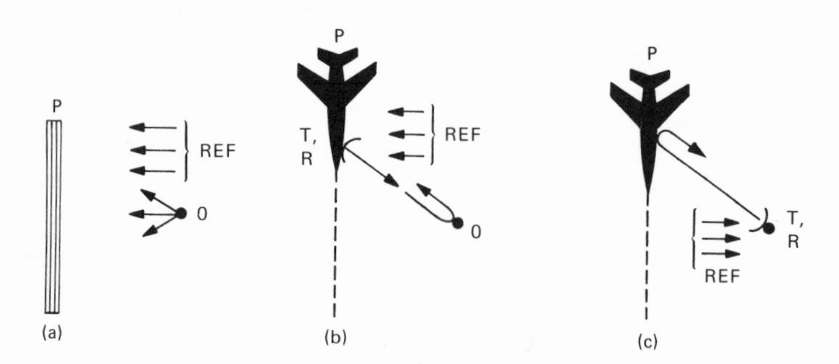

Fig. VII-11. Through the use of the comparable hologram and synthetic-aperture radar procedures, the *stationary* radar antenna at the far right can be endowed with "synthetic gain" against uniformly moving targets such as the aircraft shown.

Synthetic Gain for Stationary Radars

As we have seen, the usual synthetic-aperture radar operates from an aircraft moving in a straight line with constant speed. Through the use of a highly coherent radiated signal, it achieves the effect of a very long (synthetic-aperture) antenna, with its accompanying high gain and high resolution. We also saw, in Fig. VII-10, that a similar concept consisting of a radio transmission link can be applied to a "ground-to-air" radar. Similarly, a small ground-based, transmit–receive antenna can be effectively provided with "synthetic gain" against aircraft targets moving at constant speed along a straight line.

This process is closely related to a procedure described in 1956 by British scientist and hologram expert G. L. Rogers, and is sketched in Fig. VII-11. Again, as in Fig. VII-10, the two examples on the left portray, first, at (a) the hologram of a point source of light being recorded on the photographic plate *P*, and, second, at (b), the recording, in airborne synthetic-aperture radar, in combination with the reference signal *R*, of the signals returning from a single reflecting object *O*. At (c) the radar is stationary, and the airplane (the reflector) is assumed to be moving in a straight line at constant speed (a reference signal is again combined with the reflector signal). In both cases (b) and (c), the photographically recorded combination signal will be a one-dimensional zone plate, similar to the one we saw earlier in Fig. VII-6. The zone plate at (c) can, as can the zone plate at (b), cause coherent (laser) light to be focused so that the same "synthetic-antenna gain" is available in both cases.

We saw earlier that an airborne synthetic-aperture radar could also achieve an "end-fire" gain against a target directly in its path. In a similar way, the ground-based radar of Fig. VII-11 (c), can, with coherent radiation and a reference wave, achieve a synthetic gain against a target moving at constant speed *directly toward it* [instead of being off to one side as in Fig. VII-11 (c)]. Because this synthetic gain would only be generated against *moving* targets, reflected signals from stationary ground targets (these signals are often referred to as "ground clutter") would not be enhanced in this way. Also, some radars are designed to operate "over the horizon," depending upon energy being *diffracted* into the shadow region beyond the horizon. Improved performance in this shadow region against moving targets such as low-flying aircraft or low-flying missiles (often called "cruise missiles") should be achieved in these radars with the coherent (hologram) technique just described.

EPILOGUE

When the author was an undergraduate, "co-op" student in Electrical Engineering at the University of Cincinnati, an editorial appeared in the January, 1930, issue of the Engineering College's journal, "*The Cooperative Engineer*," written by the Head of the Electrical Department, Professor A. M. Wilson. Because the author feels that the words of that editorial are as fitting today as they were then, he would like to conclude by quoting that article in its entirety, in the hope that it may provide today's scientists and engineers with the same inspiration that he received from it in 1930.

Quo Vadimus?

by Prof. A. M. Wilson

> These are dynamic days, full of color and motion. This mundane spheroid has witnessed some rather startling changes from time to time. It has seen practically an entire civilization destroyed by sheer slaughter. It has seen one of the finest civilizations, one from which this civilization of ours has received much of its inspiration, wiped out in the short space of seventy-five years, by internecine strife. But it has never before seen anything approaching the present acceleration of what we call progress. These are the days of the quick or the dead.

Let us pause a moment, in this quiet corner, to look about us. There are many good, solid substantial minds which are becoming increasingly conscious of a disagreeable sensation of being accelerated too rapidly. "Why the hurry?," they ask, and "Where are we going in such a hurry?" And no one seems to know.

The question is constantly arising as to whether what we call progress promotes human happiness. Of course, this question is as old as human thought. It is by no means peculiar to this age. But since it is a question which confronts us, let us see if there is an answer.

Philosophers seem to be about equally divided on this question. A resume of their discussions would probably leave us about where we started. As Omar Khayyam puts it:

> "Myself, when young, did eagerly frequent
> Doctor and Saint, and heard great argument
> About it and about; but evermore
> Came out by the same door wherein I went."

H. G. Wells, in his "Outline of History," says: "The very last thing a sensible man would undertake, would be to improve mankind. The aspect of mankind is that of a great experimental workshop, in which some things in every age succeed, but most things fail; and the aim of the experiments is not the happiness of the mass, but the improvement of the type."

Probably Wells and Omar cannot be classed among the world's great thinkers, but there are millions who agree with statements of this general character. At the same time, every human being is seeking happiness in one form or another; so that, if this sort of reasoning is correct, human life becomes a rather pathetic and futile problem, and the word, progress, loses much of its value.

It is worthwhile, therefore, if we can strengthen our grasp on the conviction that true progress must of necessity increase the total of human happiness. For in his heart of hearts, practically every individual, whether or not he happens to be a college student—though I believe it is especially true of college students—likes to feel that his activities promote the success and happiness of those with whom he comes in contact.

But it is necessary to confess that a clear understanding of this problem does not imply ease of solution. While Plato was pointing out very clearly the conditions which seemed to him necessary for progress, stability, and general happiness, the Greek civilization was being destroyed by civil wars. Plato found it much easier to enunciate principles than to give them practical application.

Since it does not appear that human nature has changed greatly since Plato's time, if we are justified in looking hopefully to the future, it must be because of something new which offers promise of better results than have been achieved heretofore.

This new thing in civilization is our attitude towards what is called science, and its myraid applications throughout every phase of modern life. The freedom from drudgery which science has brought us has widened the mental horizon everywhere, and has given to citizens of the United States particularly, the definite consciousness that opportunities are available to satisfy their instinctive desire for progress. And it is quite probable that the thwarting of this instinctive desire is a very definite source of unhappiness, though an elaboration of this thought is not possible here.

If we can maintain these conditions, or make them even more widespread, we shall probably find that, while we are improving the type, we are at the same time increasing the happiness of the mass.